Contents

Vibrations and waves

PRENTICE HALL INTERNATIONAL SERIES IN PHYSICS AND APPLIED PHYSICS

Series editors:

Professor Malcolm J. Cooper, *Department of Physics, University of Warwick*
John W. Mason, *Scientific and Technical Consultant*

Beynon, J. *Introductory Optics*
Gough, W., Richards, J.P.G. and Williams, R.P. *Vibrations and Waves, 2/e*
Martin, J.L. *General Relativity: A first course for physicists*

Vibrations and waves

Second edition

W. Gough, J.P.G. Richards and R.P. Williams

PRENTICE HALL

London   New York   Toronto   Sydney   Tokyo   Singapore
Madrid   Mexico City   Munich

First published 1983 by
Ellis Horwood Limited

This second edition published 1996 by
Prentice Hall International (UK) Limited
Campus 400, Maylands Avenue
Hemel Hempstead
Hertfordshire, HP2 7EZ
A division of
Simon & Schuster International Group

Typeset in 10/12 Times by
PPS, London Road, Amesbury, Wilts.

Printed and bound in Great Britain by
Redwood Books, Trowbridge, Wiltshire

Library of Congress Cataloging-in-Publication Data

Gough, W., 1939–
   Vibrations and waves / W. Gough, J.P.G. Richards, R.P. Williams.
     p.   cm.
   Originally published: Chichester [West Sussex]: E. Horwood ; New
York : Halsted Press, 1983.
   Includes bibliographical references and index.
   ISBN 0–13–451113–1
   1. Vibration. 2. Wave-motion, Theory of. I. Richards, J. P. G.
(John Philip Gerald), 1932– . II. Williams, R. P. (Rhys Philip),
1937– . II. Title.
QA935.G67 1995
531′.1133–dc20
                                 95–18274
                                    CIP

British Library Cataloguing in Publication Data

A catalogue record for this book is available from
the British Library

ISBN 0–13–451113–1

---

1 2 3 4 5   00 99 98 97 96

Preface

Most of the important branches of physics and engineering demand for their understanding a thorough knowledge of vibrating systems and wave behaviour. It is to provide a basic route to such an understanding that the authors have written this book.

The book is designed as a course text for first- and second-year students of physics at universities, but it is hoped that it will meet the needs of engineering students too. Its contents should be well within the grasp of anyone with high-school physics or mathematics. The prior mathematical background required of the reader amounts to no more than a knowledge of elementary calculus and trigonometry; virtually no previous knowledge of vibrations or waves is demanded. Key equations are boxed for ease of reference.

Because many different types of vibration and wave are susceptible to identical mathematical treatment, much of this book is concerned with those features common to them all. It is written by experimental physicists who, while recognising the usefulness of employing the language of mathematics to describe physical phenomena, seek to treat their subject from a physical point of view. Accordingly, they endeavour to ensure that mathematical analysis is there to adorn and clarify physical reality and to draw parallels among physical phenomena which would otherwise appear unrelated.

Two of the later chapters are devoted to a study of Fourier analysis. While much of the material of the book may be studied without an understanding of Fourier methods, such methods are nowadays percolating so deeply into the outlook of the physicist and engineer that they may profitably be studied at a fairly early stage in an undergraduate's career.

The rôle of the personal computer in the educative process is by now firmly established. Suitable moving displays on a computer monitor can bring to life many of the ideas involved in the subject matter of this book more vividly than is possible by the printed word alone. Accordingly, a computer disk is available by means of which the reader may experiment with some of the concepts met. This can be ordered by using the attached postcard found at the back of this book.

The authors are indebted to Dr Martin Elliott for his frequent and patient help in the preparation of the disk, to Mr Malcolm Anderson for help in preparing the diagrams, to Mrs Pamela Tyrrell for typing part of the manuscript and to the University of Wales and University College, Cardiff, for permission to reproduce examination questions.

The authors have, over the years, had stimulating – not to say formative – conversations on vibrations, waves and many other subjects with their friend and colleague, R. Gwynne Howells, to whose memory this book is affectionately dedicated.

1 Introduction

1.1 Preamble

Although in our title, the terms *vibrations* and *waves* appear with equal prominence, they are given far from equal attention in the work itself; two chapters are devoted to vibrations, seven to waves. While the theory of vibrations has been developed primarily as essential background to waves, it is hoped that the treatment of vibrations is sufficiently full to be useful in its own right. But the book is essentially about waves. We begin by seeking to explain, in narrative terms, what waves are; and by describing, also in narrative terms, some of the kinds of wave that are to be treated fully later.

1.2 What are waves?

From an early age, we are made familiar, in different contexts, with the idea of a *wave*. We have probably, at some time, disturbed one end of a clothes line and seen the disturbance moving along the line. No doubt we have thrown a pebble into a pond and observed the consequent ripple moving outwards along the surface of the water from the point of the pebble's entry. So we have, perhaps, gradually formed the notion that a wave is a disturbance of some sort that moves in a medium. At a later stage, our instructors tell us that sound and light are both waves. At first we do not find this idea at all connected with the waves mentioned previously. They, however, assure us that this is the case, the differences being that the disturbances constituting the waves are invisible or, alternatively, even if they were visible, they are propagating with speeds too great for the eye to follow.

So we arrive at the idea that waves are disturbances in some sort of medium. In the case of the clothes line, the 'medium' was the cord itself and the 'disturbance' was a displacement of part of the cord. The disturbance was carried along by forces which different parts of the cord exerted upon their neighbours. In a similar way, the surface of a pool of water is capable of carrying a disturbance along. Here the 'medium' is the water and the 'disturbance' is an upward (or downward) movement of the surface. Sound waves in air fit into much the same mould. The 'medium' is the air itself and the 'disturbance' is a displacement of a region of the air which, owing to the elastic properties of the latter, travels along through the medium.

We have thus gone a long way towards establishing the essential nature of a wave. It is a disturbance of some sort (which may not necessarily be mechanical, as we shall see later) in a medium which, because of the properties of the latter, changes either in form or position, or both, as time goes on.

Let us consider the 'medium' a little further. Although it enables a disturbance to be propagated, it itself does not move bodily along. For example the disturbance at one end of the clothes line is propagated along its length, but none of the particles of which the cord is composed actually moves from one end of the line to the other. So a wave is a means of transmitting energy from one point to another without any net transfer of matter. Moreover, it is a means of transmitting *information* from one point to another. For example, not only does the energy in a sound wave reach our ears and activate our auditory systems, but the manner in which the energy fluctuates with time is capable of interpretation by our brains, enabling us to *perceive* the originally remote signal and (for suitable signals) to make sense of it.

It may have been noticed, during the discussion on the disturbance and the medium, that nothing at all was said about light waves and other forms of radiation such as γ-rays, X-rays, ultraviolet rays, infrared rays and radio waves. The fact is that, while these waves fit generally into the picture we have drawn, they fail to do so in one important respect – they appear to require no medium whatsoever for their propagation. Indeed, although they can travel through some material media (light waves can travel through glass, for example), they are not able to do so with as great a speed as they can through a vacuum. This was a source of such great mystery to the nineteenth-century physicist, since it was so apparent to him that all other waves required media for their propagation, that the existence of a medium, pervading the whole of space, was postulated. This was named the *luminiferous ether*, and much ingenuity was exercised in tracking down its properties. Eventually, following the classic experiment of Michelson and Morley in 1887 and the subsequent development of relativity theory, it was possible to dispense with the concept of the ether; the modern physicist now not only accepts the idea of electromagnetic waves travelling through a vacuum, but also describes them in the same mathematical way as other waves.

1.3 Description of some of the waves we shall study

1.3.1 Waves in strings

In this section we shall deal briefly, and in a general way, with the different types of wave we shall meet later on in this book. We have already mentioned waves in strings, which are particularly suitable for study in that they exemplify many of the principles common to other types of wave in a way very easy to visualise. So, although these waves can hardly be said to be of central importance in physics, the principles involved certainly are.

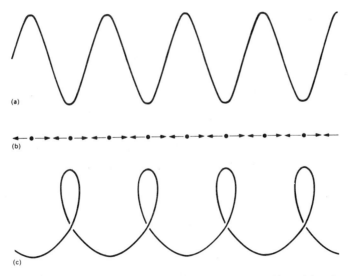

Figure 1.1 Sketches of profiles of three different waves on a string: (a) a transverse wave (displacements are vertical); (b) a longitudinal wave (arrows indicate the displacements at different parts of the string); and (c) a perspective sketch of a circularly polarised wave. The wave is basically transverse since the displacements are everywhere at right angles to the direction of propagation.

Waves in strings may be either *transverse* or *longitudinal*. For a *transverse* wave, the direction of travel (or propagation direction) is at right angles to the disturbance. Let us imagine that a string is stretched horizontally, and that one end is agitated with an up-and-down movement; then we would have a transverse wave with disturbance in the up-and-down direction moving along the length of the string. At a given instant, the string might appear as in Fig. 1.1a. On the other hand, a *longitudinal* wave can be set up by taking one end of the string and successively stretching and releasing it along its own length. In this type of wave, the disturbance is in the direction of propagation, as illustrated in Fig. 1.1b.

Transverse and longitudinal waves have many properties in common with each other, as mathematical analysis later on in this book will show. But there is one great difference. When we specify a longitudinal wave, we know that the displacements of the particles in the cord (or whatever medium is sustaining the wave) are parallel to the direction of propagation. However, in the case of a transverse wave we have to specify not only the transverse nature of the wave motion but also the direction of the transverse displacement since no particular perpendicular displacement is to be preferred to any other. For example, if the horizontally stretched clothes line is displaced in either the vertical direction or the horizontal direction perpendicular to its length, a transverse wave will result. In fact, two transverse waves with different directions of displacement can be propagated at the same time along the cord; a particularly simple wave of this sort may be obtained by rotating one end of the clothes line in a circle whose plane is perpendicular to the length of the cord. In this case a wave is produced whose profile is illustrated in Fig. 1.1c. This is usually called a *circularly polarised wave*.

1.3.2 Waves in rods

Solid rods are capable of sustaining three different types of wave. The first and simplest is the longitudinal wave which is propagated because of the tensile elastic properties of the rod. It is quite easy to excite such waves. If a rod of length about one metre is held firmly at its centre and stroked along its length by a resined cloth, a longitudinal wave is propagated back and forth along the rod. It cannot be visually observed, but the effect of the wave can easily be audibly observed as a high-pitched sound of considerable purity. Because the rod is of comparatively small length, the *reflection* of the waves back along the rod when they reach either end becomes very important and determines the pitch of the note heard. We shall be dealing in detail with the phenomenon of reflection in Chapter 5.

Transverse waves may exist in rods, but are more difficult to produce. They are mathematically much more complicated to analyse because, unlike either type of wave which exists on strings, different frequencies are propagated with different velocities. This is a phenomenon known as *dispersion* and will be studied in Chapter 6.

Finally, *torsional* waves may exist in rods. If a rod is twisted at one end, then the restoring torque causes a wave to be propagated. Here the 'disturbance' is not a linear displacement, but an angular twist. These waves are non-dispersive, that is, all frequencies are propagated at the same speed.

1.3.3 Waves on membranes

A membrane, for example the stretched skin on a drum, is the two-dimensional equivalent of a stretched string. Here, as in the string, when any part of the skin is pushed in a direction perpendicular to its plane and then released, transverse waves are set up. These waves, unlike those on the string, are two-dimensional – they spread out from the point of the initial disturbance in a way very similar to that in which surface waves on water spread out from a point of disturbance.

The simplest type of two-dimensional wave from the point of view of mathematical analysis is a 'straight-line' wave – one whose direction of propagation is constant over the whole surface and whose profile is the same along any line drawn in this direction. In general, the analysis of two-dimensional waves is not so simple, and indeed waves on a drum-head, which are reflected to the centre of the head when they reach the perimeter, require considerable sophistication for their analysis, which is outside the scope of this book.

1.3.4 Sound waves

We shall be discussing sound waves in considerable detail later on in the book. At this stage, we shall merely say that when a disturbance is created at some point, waves proceed from this point in all directions. Sound waves in gases are longitudinal because a gas is unable to sustain a torsional or shear force.

1.3.5 Waves in transmission lines

We turn back, now, to a one-dimensional example of wave motion, and one that is of great importance in certain electrical applications. An extreme convenience of electricity, as opposed to other forms of power, is that it can be conducted along metal wires to a point remote from the generator with the greatest ease. Normally, at least two parallel wires, or some similar arrangement such as coaxial wires, are required for this. There will thus be a small capacitance, and also inductance, between the wires which will cause, as we shall see in Chapter 6, a wave of voltage (and current) to travel along the line when the generator is delivering any voltage other than a perfectly steady direct current (d.c.). The result of this is that the device at the end of the line remote from the generator does not 'see' the voltage variations at the generator, but a version of these as modified by the line. This can pose problems at high frequencies, but at lower frequencies, for example the 50 Hz mains, the effects of such modifications are utterly negligible.

1.3.6 Waveguides

Point sources of waves propagating into a three-dimensional medium produce spherical waves radiating outwards from the source at the centre. This means that a detector placed at successively greater distances from the source records successively weaker responses since the total flux of energy in the wave, which is constant, is being spread over the surface of a sphere of ever-increasing surface area. This limits the effective range over which the detector can operate. In the case of sound waves, the point source would be reasonably well represented, several metres away, by a small loudspeaker, and the detector by a microphone or the ear. To increase the range it is possible to 'guide' the wave through a tube of constant internal size and shape so that the energy in the wave is not being wasted in detectorless regions. An obvious example of this arrangement, which is technically known as a *waveguide*, is the speaking tube. Other types of waveguide exist, notably the electromagnetic waveguide whose use overcomes some of the difficulties of transmitting high-frequency electrical signals along transmission lines. An important modern development in the field of telecommunications is the optical fibre through which information-carrying light waves are guided over large distances with very little loss of energy.

2 The theory of vibrations

2.1 The importance of vibration theory in the understanding of waves

When a medium is disturbed by the passage of a wave through it, the particles making up the medium are caused to vibrate. To take a simple example, corks floating on the surface of a pond will bob up and down owing to the influence of water waves. As will be seen later, the physical characteristics which describe a wave can be determined by observing the manner in which a particle in the path of the wave vibrates, so it is important at the outset to learn something of the nature and theory of vibrations.

2.2 Vibrations of a single particle

Vibrating particles are the frequent concern of writers on mechanics; the bob of a simple pendulum and the weight hanging freely from the end of a spring are obvious examples of particles which may be set in vibration, and most readers will have a good mental picture of how these vibrate. The motion is *periodic*; that is, after equal intervals of time (the *period T*) the system finds itself in exactly the same situation. The bob of the pendulum, for example, is found to be at the same position, moving with the same velocity and acceleration as it was T seconds earlier, and all these quantities will be the same T, $2T$, $3T$, etc. seconds later. Actually the swings of a pendulum die away in time, owing to frictional and viscous forces, but we are assuming an ideal pendulum which does not lose energy and goes on *ad infinitum*. During the interval of one period, a vibrating system is said to go through a *cycle* and the *frequency* (f) is defined as the number of cycles occurring in one second. Clearly, then, $f = 1/T$; the dimension of f is second^{-1}. This unit is termed the *hertz* (Hz).

2.2.1 Periodic motion of a point in one dimension

The simplest kind of periodic motion is that experienced by a point moving along a straight line, whose acceleration is directed towards a fixed point on the line and is

6

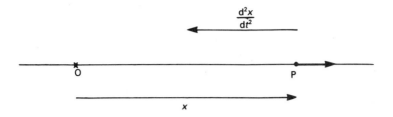

Figure 2.1 Simple harmonic motion in one dimension.

proportional to its distance from the fixed point; this is called *simple harmonic motion* (s.h.m.). Suppose a point P (Fig. 2.1) moves along a straight line so that its position with respect to a fixed point O is completely specified by the single coordinate x. Let us examine the situation when the moving point is at P (position x) and moving away from O. The acceleration is d^2x/dt^2; this is directed towards O and proportional to the distance OP $= x$. Thus

$$\frac{d^2x}{dt^2} = -\omega^2 x,$$

where ω is a constant. Note the minus sign; this is because the acceleration is directed in the opposite direction to that in which x is increasing.

On rearrangement, this equation becomes

$$\frac{d^2x}{dt^2} + \omega^2 x = 0. \tag{2.1}$$

This is a linear, second-order differential equation; linear because x and its derivative appear to the power one only, and second order because the highest derivative is d^2x/dt^2. Equation (2.1) is referred to as the differential equation governing the motion. It is not the equation of motion. To find the equation of motion we have to solve (2.1) for x. It is shown in the Appendix (A.1) that the most general solution is

$$x = a\,\sin(\omega t + \varepsilon). \tag{2.2}$$

The two arbitrary constants, inevitable in the general solution of any second-order differential equation, are, in this example, a and ε. That (2.2) is a solution of (2.1) can very easily be shown by differentiating (2.2) twice and substituting for x and d^2x/dt^2 in (2.1).

The constant a is the *amplitude* of the motion; it is the greatest possible value that x can have, since the maximum value of $\sin(\omega t + \varepsilon)$ is unity. Thus the motion takes

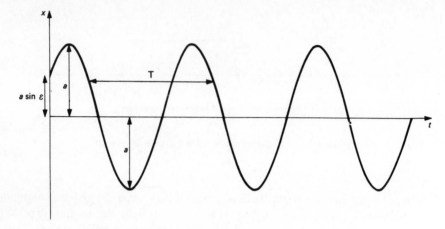

Figure 2.2   Plot of *x* against *t* for simple harmonic motion.

place entirely between the limits $x = \pm a$. The quantity $\omega t + \varepsilon$ is known as the *phase* of the motion and ε is known as the *phase constant* or *epoch*. The form of (2.2) is shown in Fig. 2.2.

Also shown in Fig. 2.2 is the period T; if we add T to t in (2.2) the value of x must remain unaltered. Hence

$$x = a \sin (\omega t + \varepsilon) = a \sin[\omega(t + T) + \varepsilon].$$

This requires the argument of the sine to be increased by exactly 2π radians. Thus by inspection, $\omega T = 2\pi$. Therefore

$$\boxed{T = 2\pi/\omega} \qquad\qquad (2.3)$$

and

$$\boxed{\omega = 2\pi f,} \qquad\qquad (2.3a)$$

since we saw earlier that the frequency $f = 1/T$. Thus we have found a physical meaning for ω, and we can write the equation of motion in physically meaningful terms as

$$x = a \sin(2\pi f t + \varepsilon). \qquad\qquad (2.4)$$

However, (2.2) is the more convenient form. ω is known as the *circular frequency* or *pulsatance* of the motion.

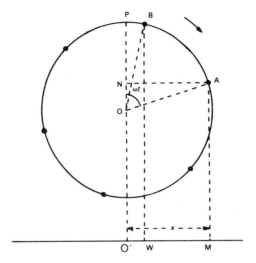

Figure 2.3 Ferris wheel.

EXAMPLE 2.1

A fairground Ferris wheel of radius 10 m has six equally spaced carriages and rotates one revolution every 20 s in a vertical plane. A man is walking along the ground, always keeping vertically below carriage A (Fig. 2.3). Show that he moves with simple harmonic motion, and deduce the amplitude, frequency, circular frequency, maximum velocity, and maximum acceleration. His wife is walking similarly below carriage B. What is her equation of motion?

Let us suppose that the clock is started (i.e. $t = 0$) when A is at its highest point P. The man is then at O'. At a later time t, his displacement x from O' is O'M, which is clearly equal to NA. But $\angle POA = \omega t$, where ω is the angular velocity of the wheel. Also, since the wheel rotates through an angle 2π in 20 s, the angular velocity is $\omega = \pi/10$ rad s^{-1}. Therefore, the equation for the man's displacement is

$$x = 10 \sin(\pi t/10). \tag{2.5}$$

By comparison with (2.4), we immediately see that the amplitude is 10 m, and the frequency is 1/20 Hz.

The circular frequency ω is $2\pi f$, which is $\pi/10$ rad s^{-1}. It should be noted that this is the same as the angular velocity of the wheel.

The velocity is obtained by differentiating (2.5), giving $dx/dt = \pi \cos \pi t/10$. This has a maximum value π m s^{-1}, this maximum occurring when the man is at O'.

Differentiating again, we obtain the acceleration $d^2x/dt^2 = -(\pi^2/10) \sin \pi t/10$, which has a maximum value of $\pi^2/10$ m s^{-2}. It should be noted that the acceleration has maximum magnitude at the *extremities* of the motion, and is zero at O', when the velocity is maximum.

The equation of motion of the man's wife W is given by the displacement $x' = 10 \sin \angle POB$. Since $\angle BOA = \pi/3$, $x' = 10 \sin(\omega t - \pi/3) = 10 \sin(\pi t/10 - \pi/3)$.

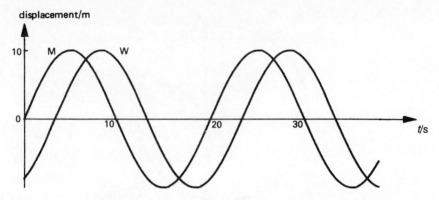

Figure 2.4   Displacement–time graphs for M and W.

The reader should appreciate that the graphs of x and x' against t (Fig. 2.4) indicate that W 'lags behind' M. The phase difference between the two is $\pi/3$, or $60°$.

This problem illustrates an important feature concerning simple harmonic motion, namely that it can be regarded as *uniform circular motion projected on to a straight line*. Indeed, some elementary texts adopt this as a definition of s.h.m. As we have seen, the angular velocity of the circular motion is equal to the circular frequency of the corresponding s.h.m. Indeed, the same symbol ω is used for both.   ■

In any given problem, the constants of integration a and ε are determined from the initial conditions. Suppose we are told that, at $t = t_0$, the point is x_0 from the origin and is moving with a velocity v_0. If we substitute these values into equation (2.4) and the first derivative of (2.4) we get

$$x_0 = a \sin(2\pi f t_0 + \varepsilon) \tag{2.6}$$

and

$$v_0 = 2\pi f a \cos(2\pi f t_0 + \varepsilon). \tag{2.7}$$

We can now solve for a and ε; first we divide (2.6) by (2.7) and invert the resulting tangent to get

$$\varepsilon = \tan^{-1}\left(\frac{2\pi f x_0}{v_0}\right) - 2\pi f t_0.$$

Secondly we square and add the equations, using the identity $\cos^2\theta + \sin^2\theta = 1$, to get

$$a^2 = x_0^2 + \frac{v_0^2}{4\pi^2 f^2}.$$

There are numerous situations in physics which can be treated to a very good approximation as involving simple harmonic motion; we shall develop just two, one mechanical and one electrical.

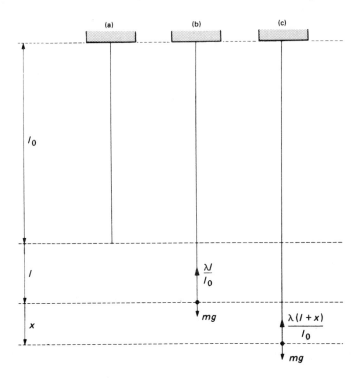

Figure 2.5 Motion of a mass suspended from a light, elastic string: (a) string hanging freely; (b) string with mass *m* suspended from end; and (c) mass in oscillation, at the instant when it is a distance *x* below equilibrium position.

Our mechanical example is that of a mass suspended from one end of a light elastic string, the other end of which is rigidly clamped (Fig. 2.5). Suppose the string is light, has an unstretched length l_0 and an elastic constant λ; when a mass *m* is hung from the end and lowered slowly to the equilibrium position, the string will be stretched a distance *l* such that, by Hooke's law,

$$mg = \frac{\lambda l}{l_0},\qquad(2.8)$$

where *g* is the acceleration due to gravity. Since λ and l_0 are both constants for a given string, we can replace λ/l_0 by the single constant *k*, which is known as the 'stiffness' or force constant of the string; so (2.8) becomes

$$mg = kl.\qquad(2.9)$$

If the mass is pulled down a small distance from the equilibrium position and released, vertical oscillations will ensue. Applying Newton's second law of motion (i.e. equating mass times acceleration to net force) to the situation shown in Fig. 2.5c, when the mass is instantaneously a distance x below the equilibrium position, we have

$$m \frac{d^2 x}{dt^2} = mg - k(l + x).$$

Note that x increases in the downward direction, and is therefore a positive quantity.
    From (2.9)

$$m \frac{d^2 x}{dt^2} = -kx$$

or

$$m \frac{d^2 x}{dt^2} + kx = 0. \tag{2.10}$$

This has exactly the same form as equation (2.1), with $\omega^2 = k/m$, so we can write down the general solution of (2.10) as

$$x = a \sin\left[\sqrt{\left(\frac{k}{m}\right)} t + \varepsilon\right], \tag{2.10a}$$

which is the equation of motion of the mass. Thus the mass performs simple harmonic oscillations with period $2\pi/\omega$, i.e.

$$T = 2\pi \sqrt{\left(\frac{m}{k}\right)};$$

the constants a and ε could be determined from initial conditions as explained earlier.
    The electrical example is that of the circuit shown in Fig. 2.6 which comprises a capacitor C in series with a coil of pure inductance L. The capacitor is charged from a battery, which is then removed, and the key K closed. The capacitor will thereupon

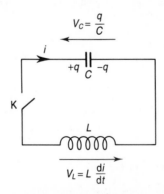

Figure 2.6   Circuit containing inductance and capacitance.

discharge through the coil. Suppose, at some instant subsequent to the closing of the key, the charges on the plates of the capacitor are $\pm q$ and the current i is flowing in the direction indicated. The potential differences across C and L will have the values shown, the arrows indicating the direction of increasing potential. The algebraic sum of the potential differences round a closed circuit must be zero at any instant, so

$$L\frac{di}{dt} + \frac{q}{C} = 0.$$

But

$$i = \frac{dq}{dt},$$

so we have

$$L\frac{d^2q}{dt^2} + \frac{q}{C} = 0. \tag{2.11}$$

The charge on the capacitor thus fluctuates with time according to the solution of (2.11), which is

$$q = q_0 \sin\left(\frac{1}{\sqrt{(LC)}} t + \varepsilon\right). \tag{2.12}$$

As before, q_0 and ε could be determined from the initial conditions.

In both the above examples, the oscillations continue indefinitely with constant amplitude. In real systems the oscillations eventually die away, and this will be investigated fully in Chapter 3. The reason the oscillations continue indefinitely here is that no energy leaves the system, and we can make use of this fact in deducing (2.10) by the following alternative method. If we define the potential energy (PE) of the system of Fig. 2.5 to be zero when the mass is at the equilibrium level, then when the displacement is x the PE is $\frac{1}{2}kx^2$ and the kinetic energy (KE) is $\frac{1}{2}m(dx/dt)^2$. The total energy $U = \mathrm{KE} + \mathrm{PE}$ does not change with time, so its derivative is zero. Thus

$$\frac{dU}{dt} = \frac{d}{dt}\left\{\tfrac{1}{2}kx^2 + \tfrac{1}{2}m\left(\frac{dx}{dt}\right)^2\right\} = 0$$

so

$$kx\left(\frac{dx}{dt}\right) + m\left(\frac{dx}{dt}\right)\left(\frac{d^2x}{dt^2}\right) = 0$$

giving $m(d^2x/dt) + kx = 0$ as before.

Similarly, for the circuit of Fig. 2.6, the magnetic PE stored by the inductor is $\frac{1}{2}Li^2$, and the electrostatic PE stored by the capacitor is $\frac{1}{2}(q^2/C)$. The sum of these, which is the total electromagnetic energy of the circuit U, is constant. Equating dU/dt to zero and substituting $i = dq/dt$ quickly leads to (2.11).

2.2.2 Periodic motion of a point in two dimensions

Suppose a point is subjected simultaneously to two simple harmonic motions at right angles to each other. A physical example of this is the motion of the spot on a cathode-ray oscilloscope screen when alternating voltages are applied simultaneously to the X- and Y-plates.

Suppose that the voltage across the X-plates causes the spot to move according to the equation

$$x = a \sin 2\pi ft, \tag{2.13}$$

and that across the Y-plates causes the spot to move according to the equation

$$y = b \sin(2\pi ft + \varepsilon). \tag{2.14}$$

This means that the alternating voltages have the same frequency f, different amplitudes, and differ in phase by ε radians. When the voltages are simultaneously applied, we can find the path described by the spot by eliminating the time t from equations (2.13) and (2.14) as follows. Expanding (2.14) we get

$$y = b \sin 2\pi ft \cos \varepsilon + b \cos 2\pi ft \sin \varepsilon. \tag{2.15}$$

From equation (2.13) we see that

$$\sin 2\pi ft = \frac{x}{a},$$

so that

$$\cos 2\pi ft = \sqrt{\left(1 - \frac{x^2}{a^2}\right)}.$$

Inserting these values into (2.15) and simplifying, we obtain

$$a^2 y^2 + b^2 x^2 - 2abxy \cos \varepsilon - a^2 b^2 \sin^2 \varepsilon = 0. \tag{2.16}$$

Readers may recognise this as the general equation for an ellipse. The form can readily be determined, and the curve sketched, by putting specific values of x and y into (2.16).

$$
\begin{aligned}
\text{When } y = 0 \quad & x = \pm a \sin \varepsilon, \\
x = 0 \quad & y = \pm b \sin \varepsilon, \\
x = \pm a \quad & y = \pm b \cos \varepsilon, \\
y = \pm b \quad & x = \pm a \cos \varepsilon.
\end{aligned}
$$

These points are plotted to give the curve shown in Fig. 2.7.

Thus the picture we see on the screen will, in general, be an ellipse, and it is worth noting that the phase angle ε can be determined from it since

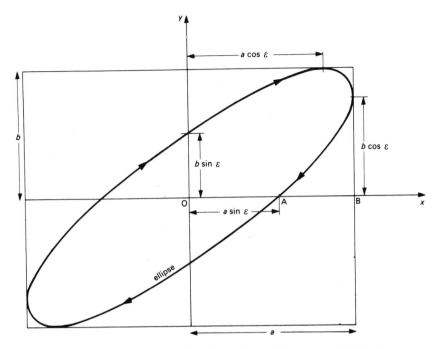

Figure 2.7 Two simple harmonic motions, of equal frequency, at right angles.

$$\frac{OA}{OB} = \sin \varepsilon.$$

For special values of ε ($\varepsilon = 0$ and π) the ellipse degenerates into a straight line, whilst for $\varepsilon = \frac{1}{2}\pi$ and $\frac{3}{2}\pi$ the major and minor axes of the ellipse coincide with the x- and y-axes. These particular instances are shown in Fig. 2.8. The situation shown in Fig. 2.8a could have easily been predicted; since the x- and y-motions are in phase with one another, the maximum of x will occur at the same instant as the maximum of y. The situation shown in Fig. 2.8b for $\varepsilon = \frac{1}{2}\pi$ (and Fig. 2.8d for $\frac{3}{2}\pi$) is also of interest. If $a = b$ we have a circular path; this point will be pursued further during the discussion of polarisation in Chapter 10.

The curves in Fig. 2.8 are examples of Lissajous figures. If the frequencies of the two signals are not the same, the figures become more complicated and are not, in general, stationary except when the frequencies are in the ratio of two whole numbers.

2.3 Coupled vibrators

We now need to see how vibrational behaviour changes when the simple vibrating systems of the previous section are connected together or 'coupled'.

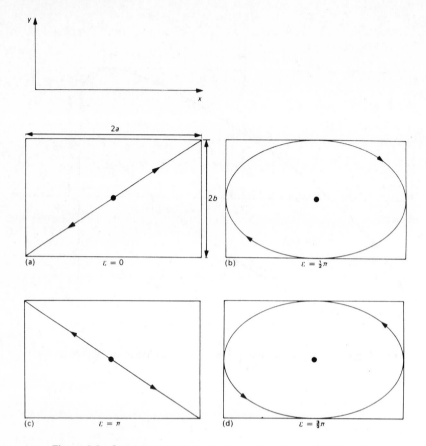

Figure 2.8 Special cases of simple harmonic motions at right angles.

2.3.1 Vibrations of two coupled particles: normal modes

We consider first mechanical systems consisting of two particles coupled together. Examples of the types of system we have in mind are shown in Fig. 2.9. Eventually we shall obtain equations of motion of the masses shown in Fig. 2.9c, but before we embark on the mathematics it is very important that readers should obtain a clear picture of the ways in which the particles in such systems can vibrate, so we strongly recommend that anyone meeting this subject for the first time should try – or at least imagine – the following simple experiment. The apparatus required is a 'slinky' spring (obtainable from a toy shop) and about a metre of the type of chain used to secure wash-basin stoppers. The spring is cut into halves; the upper ends are clamped at the same level, as shown in Fig. 2.10, and the ends of the chain are connected to the lower ends of the springs.

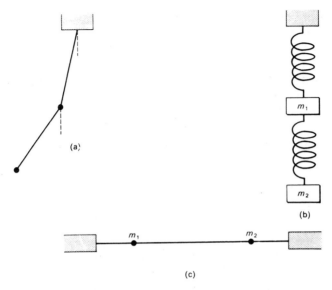

Figure 2.9 Systems with coupled particles: (a) double pendulum; (b) two suspended masses and springs; and (c) stretched elastic string, loaded with two masses.

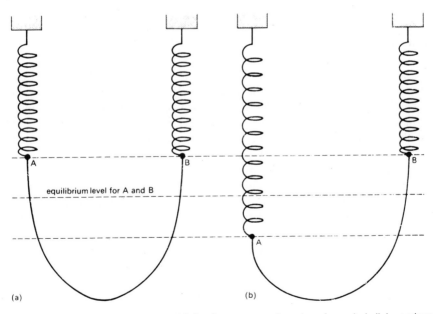

Figure 2.10 Starting positions of A and B for the two normal modes of coupled slinky springs: (a) mode 1 and (b) mode 2.

The system so obtained is rather different from those shown in Fig. 2.9, since it has a distributed mass rather than point masses. This makes the analysis much harder, but this does not matter since we do not intend to treat this system mathematically. We want to note how the positions of the points A and B, where the chain joins the springs, vary with time when the system is set into vertical motion.

First, raise A and B by equal amounts and release them from rest. A and B should be seen to oscillate up and down with simple harmonic motion. Clearly A and B are exactly in phase – they are both at the top of the motion (or both at the bottom) at the same instant. Find an approximate value for the period. We will refer to this particular motion as mode 1.

Next lower A and raise B by equal amounts to the configuration shown in Fig. 2.10b and release them from rest. Once more the points A and B should be seen to perform simple harmonic motion, but now in exact antiphase (A is at the top of its motion whilst B is at the bottom, and vice versa). We will refer to this motion as mode 2. Note that the period is not the same as for mode 1.

Bring the system to rest once more, and start motion by drawing A, alone, down and releasing it. The motion which follows (mode 3) should be seen to be of a much more complicated character than previously; neither A nor B oscillates with simple harmonic motion, but the variation of the position of each with time should be somewhat as shown in Fig. 2.11. Note that the oscillations of both points successively build up and die away, and that A is oscillating with its greatest amplitude when B has minimum amplitude, and vice versa. The kinetic energy of the system is, in fact, being periodically transferred from one side to the other.

Modes 1 and 2 are known as *normal modes*; a system of two particles is said to be in a normal mode when both particles are executing simple harmonic motion with the same frequency. Mode 3 is a general state of motion. Note that there are two distinct normal modes for this system, and that they have different frequencies associated with them. A system whose configuration is completely described by two coordinates, such as the positions of the points A and B in our example, is said to have two *degrees of freedom*. Such a system has two distinct normal modes of vibration. Note that it is incorrect in general to equate the number of degrees of freedom to the number of particles, since any particle may have up to three degrees of freedom. In our example the points A and B move up and down only, so the position of each is completely specified by a single coordinate.

Now that the reader has a qualitative idea of the way in which a system with two particles vibrates, the time has come to provide a proper quantitative analysis of the problem. We will take the system of Fig. 2.9c, a stretched light elastic string of length $3a$ fixed at both ends and loaded with two equal masses at the points of trisection; it will be seen later why this choice is an appropriate one for a book on waves. Let us assume that such a system has been set in motion, and that at a given instant the two masses are displaced transversely, as shown in Fig. 2.12.

The following assumptions are made:

1. The displacements x_1 and x_2 of the masses from their equilibrium positions,

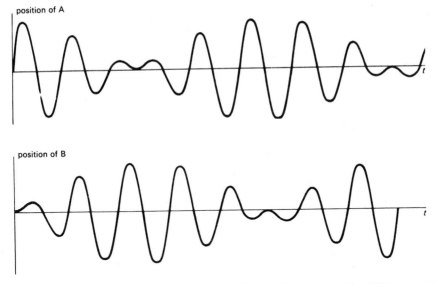

position of A

position of B

Figure 2.11 Graphs of position against time for A and B when the system of Fig. 2.10 is in a general state of motion.

which are exaggerated for clarity in Fig. 2.12, are always small in comparison with the length of the string.

2. The changes in the original tension T due to these small displacements can be ignored in comparison with T itself.

3. The effects of gravity are ignored.

To obtain the differential equations governing the motions of the particles, we apply Newton's second law of motion. For the first mass,

$$m \frac{d^2 x_1}{dt^2} = T \sin \theta_2 - T \sin \theta_1,$$

which, when we replace $d^2 x_1/dt^2$ by $\ddot{x}_1$ for brevity, becomes

$$m\ddot{x}_1 = T \sin \theta_2 - T \sin \theta_1, \tag{2.17}$$

and for the second mass

$$m\ddot{x}_2 = -T \sin \theta_2 - T \sin \theta_3; \tag{2.18}$$

where the angles $\theta_1, \theta_2, \theta_3$ are those shown in Fig. 2.12.

Since x_1 is small, $\sin \theta_1 \cong \tan \theta_1 = x_1/a$; similarly $\sin \theta_2 \cong (x_2 - x_1)/a$ and $\sin \theta_3 \cong x_2/a$. Inserting these values into (2.17) and (2.18), and rearranging, we obtain

$$m\ddot{x}_1 + \frac{2Tx_1}{a} - \frac{Tx_2}{a} = 0, \tag{2.19}$$

$$m\ddot{x}_2 + \frac{2Tx_2}{a} - \frac{Tx_1}{a} = 0. \tag{2.20}$$

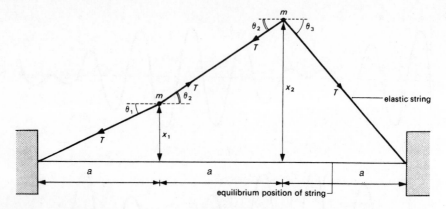

Figure 2.12   Loaded elastic string at some instant during motion.

Equations (2.19) and (2.20) are simultaneous differential equations. Rather than solve them directly, we will investigate the conditions under which the system will oscillate in a normal mode. We have already seen, from the experiment, that in a normal mode the particles execute simple harmonic motion with the same frequency; we can therefore write down the equations of motion for the two masses as

$$x_1 = A \sin \omega t, \tag{2.21}$$

$$x_2 = B \sin \omega t. \tag{2.22}$$

In the slinky spring experiment, the reference points A and B moved either in phase or in antiphase during normal vibrations. This is generally true. There is no need therefore to include a phase constant (ε in (2.2)) since the particles will either be exactly in phase or exactly in antiphase; in the latter case, A and B will have opposite signs because

$$\sin(\omega t + \pi) = -\sin \omega t.$$

Substituting from (2.21) and (2.22) into (2.19) and (2.20) we obtain, after cancelling out the trigonometric terms,

$$-m\omega^2 A + \frac{2TA}{a} - \frac{TB}{a} = 0 \tag{2.23}$$

and

$$-m\omega^2 B + \frac{2TB}{a} - \frac{TA}{a} = 0. \tag{2.24}$$

We can eliminate the ratio A/B between these equations to get

$$\left[-m\omega^2 + \frac{2T}{a} \right]^2 = \frac{T^2}{a^2};$$

so

$$-m\omega^2 + \frac{2T}{a} = \pm\left(\frac{T}{a}\right).$$

Thus, solving for ω^2 we have

$$\omega^2 = \frac{T}{ma} \quad \text{or} \quad \frac{3T}{ma}.$$

Thus there are two different circular frequencies with which this system oscillates in a normal mode; these are

$$\omega_1 = \sqrt{\left(\frac{T}{ma}\right)} \tag{2.25}$$

and

$$\omega_2 = \sqrt{\left(\frac{3T}{ma}\right)}. \tag{2.26}$$

If we now substitute the value of ω_1 into (2.23) we get

$$A = B.$$

Thus, in this (the first) normal mode, the amplitudes of motion of the masses are the same, and the latter move in phase at all times; the vibration will therefore be as shown in Fig. 2.13a.

If we substitute ω_2 into (2.23) we get $A = -B$ in the second normal mode, therefore, the amplitudes are the same but the motions are in antiphase, as indicated in

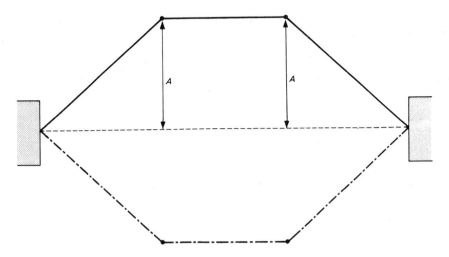

Figure 2.13a The first normal mode.

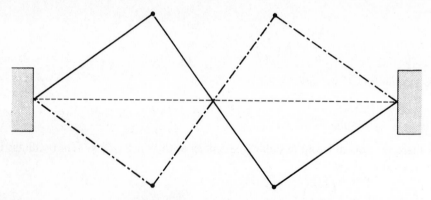

Figure 2.13b   The second normal mode.

Fig. 2.13b. The frequencies f_1 and f_2 corresponding to the normal modes are referred to as the *natural frequencies* of the system. (Remember $f = \omega/2\pi$.)

We must now find the equations of motion of the masses when the system is in a general state of motion. Knowledge that in the normal modes $x_1 = x_2$ or $x_1 = -x_2$ at all times suggests how we should proceed. If we add together equations (2.19) and (2.20) we obtain

$$m(\ddot{x}_1 + \ddot{x}_2) + \frac{2T}{a}(x_1 + x_2) - \frac{T}{a}(x_1 + x_2) = 0, \tag{2.27}$$

and if we subtract them we get

$$m(\ddot{x}_1 - \ddot{x}_2) + \frac{2T}{a}(x_1 - x_2) + \frac{T}{a}(x_1 - x_2) = 0. \tag{2.28}$$

Writing $x_1 + x_2 = u$ in (2.27), and $x_1 - x_2 = v$ in (2.28), so that $\ddot{x}_1 + \ddot{x}_2 = \ddot{u}$ and $\ddot{x}_1 - \ddot{x}_2 = \ddot{v}$, we find that these equations simplify to

$$m\ddot{u} + \frac{T}{a}u = 0,$$

$$m\ddot{v} + \frac{3T}{a}v = 0.$$

These we recognise as the differential equations for simple harmonic motion (see (2.1)), so we can immediately write down their solutions (see (2.2)) as follows:

$$u = x_1 + x_2 = C \sin\left[\sqrt{\left(\frac{T}{ma}\right)}t + \alpha\right]$$

and

$$v = x_1 - x_2 = D \sin\left[\sqrt{\left(\frac{3T}{ma}\right)}t + \beta\right],$$

where C, D, α and β are the arbitrary constants of integration. We can solve for x_1 and x_2 as follows:

$$x_1 = \frac{u + v}{2}$$

$$= \tfrac{1}{2}C \sin\left[\sqrt{\left(\frac{T}{ma}\right)}t + \alpha\right] + \tfrac{1}{2}D \sin\left[\sqrt{\left(\frac{3T}{ma}\right)}t + \beta\right] \qquad (2.29)$$

and

$$x_2 = \frac{u - v}{2}$$

$$= \tfrac{1}{2}C \sin\left[\sqrt{\left(\frac{T}{ma}\right)}t + \alpha\right] - \tfrac{1}{2}D \sin\left[\sqrt{\left(\frac{3T}{ma}\right)}t + \beta\right]. \qquad (2.30)$$

Equations (2.29) and (2.30) are the equations of motion of the masses when the system is in a general state of motion. It will be seen that the motion in each case is the sum of two simple harmonic terms, and that the frequencies of these simple harmonic terms are, respectively, the frequencies associated with the two normal modes in which the system is capable of oscillating. This last point can be checked by comparing the circular frequency terms in (2.29) and (2.30) with ω_1 and ω_2 in equations (2.25) and (2.26). Mathematically, equations (2.29) and (2.30) are said to be *linear combinations* of the normal mode solutions.

The type of motion obtained when a particle executes two simple harmonic motions of different frequencies simultaneously is known as *beats*. This is the effect shown in Fig. 2.11. At certain times the two motions will be in step with one another, resulting in a large displacement; but since the frequencies are different they will gradually get out of step and will eventually tend to cancel each other out, then later they will get back into step and so on.

The phenomenon of beats is more usually associated with the combination of two collinear simple harmonic motions of nearly the same frequency. Suppose we have two such motions of the same amplitude a, but whose circular frequencies differ by a small amount $\Delta\omega$. On combining, we get

$$y = a \sin \omega t + a \sin(\omega + \Delta\omega)t.$$

Applying the trigonometrical formula

$$\sin A + \sin B = 2 \sin \tfrac{1}{2}(A + B) \cos \tfrac{1}{2}(A - B),$$

we have

$$y = 2a \cos\left[\frac{\Delta\omega}{2}t\right] \sin\left[\left(\omega + \frac{\Delta\omega}{2}\right)t\right].$$

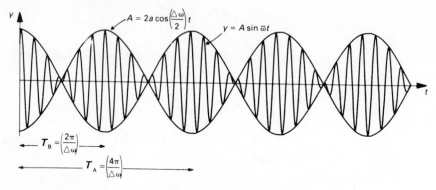

Figure 2.14 Beats.

This can be written as

$$y = A \sin \bar{\omega}t,$$

where $A = 2a \cos((\Delta\omega/2)t)$ and $\bar{\omega}$ is the mean of the two original circular frequencies. The result is therefore similar to simple harmonic motion, except that the amplitude A changes slowly with time.

Probably the best example of this is the note produced when two tuning forks of the same nominal frequency are struck simultaneously. This note is heard to build up and die away alternately with low frequency. Since the ear recognises intensity but not phase, it is unable to distinguish between the maxima and minima of A, so the apparent frequency of the beats (the so-called beat frequency) is $\Delta f = \Delta\omega/2\pi$ (the difference between the individual frequencies of the forks).

Figure 2.14 shows how the resultant disturbance y varies with time. It can be seen that the beat period $T_B = 2\pi/\Delta\omega$ is exactly one half of the period T_A.

2.3.2 Coupled electrical circuits

For our example of coupled electrical circuits, we consider two resistance-free LC circuits, identical to that in Fig. 2.6, arranged so that the two inductors are close together as shown in Fig. 2.15. Thus the changing magnetic field in the first coil induces an electromotive force (e.m.f.), and hence a current, in the second coil, and vice versa. This process is known as mutual induction and the coefficient of mutual induction, M, is defined by $\varepsilon_2 = M(di_1/dt)$, where ε_2 is the e.m.f. induced in the second coil due to the changing current i_1 in the first. Similarly $\varepsilon_1 = M(di_2/dt)$.

As in section 2.2 we set the system into oscillation, for example by charging the left-hand capacitor – the right-hand capacitor is initially uncharged – and then closing the key K. The various potentials and currents a time t later are shown in Fig. 2.15 from which it will be noted that those for the left-hand circuit are the same as in

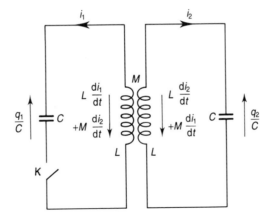

Figure 2.15 Two identical *LC* circuits coupled through mutual inductance.

Fig. 2.6 but with the mutual induction term $M(di_2/dt)$ now included. Summing the potential differences round each circuit and equating to zero, as before, we have

$$L\frac{di_1}{dt} + M\frac{di_2}{dt} + \frac{q_1}{C} = 0$$

$$L\frac{di_2}{dt} + M\frac{di_1}{dt} + \frac{q_2}{C} = 0.$$

But now $i_1 = dq_1/dt$ and $i_2 = dq_2/dt$. Inserting these in the above equations we have:

$$L\frac{d^2q_1}{dt^2} + M\frac{d^2q_2}{dt^2} + \frac{q_1}{C} = 0$$

$$L\frac{d^2q_2}{dt^2} + M\frac{d^2q_1}{dt^2} + \frac{q_2}{C} = 0.$$

To solve these, we proceed as in section 2.3.1 for the string loaded with two masses (Fig. 2.12). Adding the above two equations and writing $d^2q_1/dt^2 = \ddot{q}_1$, etc. we have

$$L(\ddot{q}_1 + \ddot{q}_2) + M(\ddot{q}_1 + \ddot{q}_2) + \frac{1}{C}(q_1 + q_2) = 0.$$

And subtracting

$$L(\ddot{q}_1 - \ddot{q}_2) - M(\ddot{q}_1 - \ddot{q}_2) + \frac{1}{C}(q_1 - q_2) = 0.$$

Note that these are similar to, but not quite analogous with, (2.27) and (2.28).
Writing $u = q_1 + q_2$, $v = q_1 - q_2$ we obtain

$$(L + M)\ddot{u} + \frac{1}{C}\, u = 0$$

$$(L - M)\ddot{v} + \frac{1}{C}\, v = 0.$$

These are, of course, s.h.m. equations and have solutions:

$$u = 2A \,\sin(\omega_1 t + \alpha)$$

$$v = 2B \,\sin(\omega_2 t + \beta)$$

where

$$\omega_1^2 = \frac{1}{C(L + M)},$$

and

$$\omega_2^2 = \frac{1}{C(L - M)}.$$

We have included the 2 in the arbitrary constants for later convenience.

If we write $\omega_0^2 = 1/LC$ for the single isolated circuit, we have

$$\omega_1 = \frac{\omega_0}{\sqrt{\left(1 + \dfrac{M}{L}\right)}},$$

and

$$\omega_2 = \frac{\omega_0}{\sqrt{\left(1 - \dfrac{M}{L}\right)}}.$$

Solving the above equations for u and v to obtain the time variation of charge on each capacitor we obtain

$$q_1 = \frac{u + v}{2} = A \,\sin(\omega_1 t + \alpha) + B \,\sin(\omega_2 t + \beta) \tag{2.31}$$

$$q_2 = \frac{u - v}{2} = A \,\sin(\omega_1 t + \alpha) - B \,\sin(\omega_2 t + \beta) \tag{2.32}$$

which correspond to (2.29) and (2.30) for the loaded string.

Thus in the general case we have beats (see Fig. 2.14). This would for example be the outcome if the circuits were set oscillating as described at the start of this section: viz. at $t = 0$, $q_1 = Q$, $q_2 = \dot{q}_1 = \dot{q}_2 = 0$. But the algebra is tedious and the end result of little interest, so we will not pursue it.

Of more interest are the normal mode solutions which occur when both circuits oscillate with the same frequency and are either in phase or in antiphase. These can be identified readily from (2.31) and (2.32) above. When $B = 0$ we have $q_1 = q_2 = A \sin(\omega_1 t + \alpha)$. This is the first (in-phase) normal mode and can be obtained in practice by giving each capacitor the same charge $+Q$ at $t = 0$ at which time we complete both circuits so the initial currents are both zero. By inserting these initial conditions we find that $A = Q$, $\alpha = \pi/2$ and therefore $q_1 = q_2 = Q \cos(\omega_1 t)$.

The second (antiphase) normal mode is obtained when $A = 0$ so that, from (2.31) and (2.32) we see that

$$q_1 = -q_2 = B \sin(\omega_2 t + \beta).$$

This normal mode is set up by giving the capacitors equal but opposite charges $\pm Q$ at $t = 0$.

If the initial currents are both assumed zero as before we find $\beta = \pi/2$ and $q_1 = -q_2 = Q \cos(\omega_2 t)$. As in the case of the loaded string we see that the antiphase normal mode has the higher frequency because $\omega_2 > \omega_1$.

We have chosen the simplest possible case. The general treatment, with all circuit components different and arbitrary starting conditions, involves heavy algebra and gives no better insight.

2.4 Systems with more than two particles

Systems of more than two particles can be analysed along the lines of section 2.3, but the mathematics becomes rather cumbersome. A system comprising a stretched elastic string loaded with three evenly spaced equal masses can be shown to have three normal modes of vibration and three natural frequencies. The normal modes are shown in Fig. 2.16. For this system, the general equation of motion for any of the particles can also be shown to be a linear combination of the three normal mode solutions.

We can extend the pattern further. A system of N particles (each particle having one degree of freedom only) can be shown to have N normal modes of vibration and N natural frequencies, and the general state of motion is a linear combination of the N normal mode solutions.

An important extension of these ideas is encountered in the vibrations of a stretched heavy string. Such a string can be regarded as a system comprising an infinite number of particles, and will have an infinite number of normal modes and natural frequencies. It is possible to analyse the behaviour of a vibrating string in this way, but it is much more convenient to regard the string as a continuous medium, and to treat the problem by means of waves, as is done in Chapter 6.

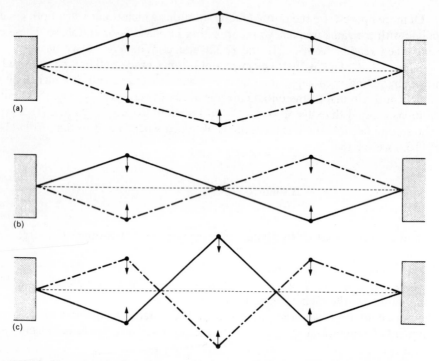

Figure 2.16 Normal modes for a three-particle system: (a) mode 1 (lowest frequency); (b) mode 2 (note that the central mass is permanently at rest); and (c) mode 3 (highest frequency).

2.5 Non-linear vibrations

We consider in this section vibrations where the restoring force is no longer proportional to the displacement, for example a simple pendulum oscillating with large amplitude, a mass suspended from a spring which does not obey Hooke's law or the vertical oscillations of a floating spherical ball. We will look at the floating ball in detail.

Suppose the ball has radius a and density $\rho/2$, and floats in water of density ρ. Then according to Archimedes' principle the ball will float exactly half submerged since the upthrust, which is equal to the weight of water displaced, is exactly equal to the weight of the ball.

If the ball is now pushed down a vertical distance x into the liquid, as shown in Fig. 2.17, a further volume of liquid is displaced which we can find by volume integration – adding up all the elementary discs of radius $r = \sqrt{(a^2 - z^2)}$ from $z = 0$ to $z = x$, i.e.

$$\int_0^x \pi(a^2 - z^2)\, \mathrm{d}z = \pi(a^2 x - x^3/3).$$

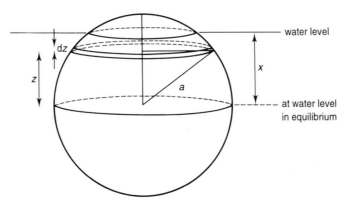

Figure 2.17 Floating sphere. The sphere floats exactly half submerged in equilibrium, but when depressed a distance x down into the water, an upthrust is produced which is equal to the weight of water displaced as a consequence.

The additional upthrust, equal to the weight of extra water displaced, is therefore

$$\pi(a^2x - x^3/3)\rho g.$$

Finally if the ball is released and we apply Newton's second law of motion, equating mass × acceleration to the restoring force in the now-familiar way, we obtain, writing the mass of the sphere in terms of its volume and density (which is $\rho/2$)

$$\frac{4}{3}\pi a^3 \frac{\rho}{2} \ddot{x} = -\pi(a^2x - x^3/3)\rho g$$

or, more tidily,

$$\ddot{x} + \frac{3g}{2a}\left(x - \frac{x^3}{3a^2}\right) = 0. \tag{2.33}$$

Clearly the restoring force is not proportional to x and the resulting vibrations are not s.h.m.

Equation (2.33) is not easy to integrate, and we shall not attempt it. If we make the usual assumption of no loss of energy, the subsequent motion is bound to be periodic in that we will have the usual transference back and forth between kinetic and potential energies, but we shall find that the period is no longer constant but dependent upon the initial conditions. This can be seen qualitatively by the following argument. If we consider first small oscillations for which $x \ll a$ we see that the term in x^3 can be neglected, and (2.33) reduces to

$$\ddot{x} + \frac{3g}{2a}x = 0.$$

This we recognise as s.h.m. with

$$\omega^2 = \frac{3g}{2a}$$

and therefore period

$$T = 2\pi \sqrt{\left(\frac{2a}{3g}\right)}.$$

As x is increased, we see that the restoring force is diminished due to the x^3 term. Thus the acceleration is diminished and the time taken for a cycle is increased compared with the s.h.m. situation. It follows that the period increases as the initial displacement increases; it is therefore not constant but dependent on the amplitude. The more we press the ball in to start with, the greater will be the period. Differential equations such as (2.33) are difficult to solve, and often can only be tackled by numerical (as opposed to analytical) methods. Another example of non-linear vibrations is that of the simple pendulum when the angle of oscillation is large so that the approximation $\sin \theta = \theta$ is no longer valid. The resulting differential equation is one which can be solved analytically, but it is difficult. The reader may verify that the period increases with amplitude by an argument similar to that given above, remembering that $\sin \theta \approx \theta - \theta^3/6$.

Perhaps the most striking example of non-linear vibrations is the *frequency doubling* which occurs when strong laser light passes through certain crystals. The oscillating electric field of the incident light causes oscillating *polarisation* (the separation of positive and negative charges) in the constituent molecules of the crystal. The resulting oscillating charges act as sources for re-emitted light. The relation between the applied electric field E and the resulting polarisation P is

$$P = \varepsilon_0\{\chi E + \chi' E^2 + \chi'' E^3 + \ldots\} \tag{2.34}$$

where ε_0 is the permittivity of free space and, χ, χ' and χ'' are the susceptibilities of the crystal. Equation (2.34) is clearly non-linear on account of the terms in E^2 and above, but, for typical fields E, the magnitudes of the terms in (2.34) diminish rapidly as we proceed to the right.

We can most conveniently see the effect of an oscillating field by writing E in complex exponential form as described in detail in section 4.3. This, as we shall see, enables us to express the field oscillating with circular frequency ω, viz.

$$E = E_0 \sin \omega t;$$

in the mathematically more convenient form

$$E = E_0 \exp(i\omega t), \tag{2.35}$$

where exp is the exponential operator and $i = \sqrt{(-1)}$.

Inserting (2.35) into (2.34) we obtain

$$P = \varepsilon_0\{\chi E_0 \exp(i\omega t) + \chi' E_0^2 \exp(i2\omega t) + \chi'' E_0^3(i3\omega t) + \ldots\} \tag{2.36}$$

In the case of non-laser incident light, the value of E_0 is low and only the first term on the right is of any significant size. This is the uninteresting linear case in which the re-emitted light has the same circular frequency, ω, and hence the same wavelength as the incident light. However when laser light of high E_0 is incident, the second term becomes significant and we can see that this has a circular frequency 2ω. Hence the term frequency doubling. It turns out that only piezoelectric crystals (see section 3.6) show frequency doubling to a readily observable extent, as only they have the required crystal symmetry to produce non-zero χ'. A simple experiment to demonstrate frequency doubling consists of directing laser light through a crystal of, say, potassium dihydrogen phosphate and thence onto a triangular prism to resolve the resulting spectrum into two lines – a strong one of circular frequency ω and a much weaker one of circular frequency 2ω (i.e. half the wavelength).

2.6 Chaotic vibrations

A development of non-linear vibrations which is important and peculiarly interesting is the topic of chaotic vibrations, or, more generally, chaos in dynamic systems. The basic idea behind chaos can be appreciated through a simple experimental example – that of a dripping water tap. If a tap drips slowly, as it will when the valve is almost fully closed, the time between the descent of successive drops turns out to be constant. A graph of this would consist of equally spaced blips along the time axis. If the valve is opened up gradually there comes a point when the time between drops is no longer constant. If the successive times are recorded over a long period and plotted, it is found that there is no *short-range order* (times between successive drops are not constant) or *long-range order* (the sequence of times do not appear to repeat themselves even over a long period of time). This complete lack of order is an example of what has come to be known as chaos. Chaos is perhaps not a good term, for the behaviour of a dripping tap can in fact be predicted from the laws of physics and a knowledge of the initial conditions – very difficult, but in principle possible. In other words the future pattern of events is completely determined by the starting conditions. What actually happens is that minute differences in the starting conditions result in major differences in subsequent behaviour. Thus, in the so-called butterfly effect, the presence of a butterfly is considered sufficient to alter the starting conditions so as to change the subsequent weather from what would otherwise have been calm into violent storm. This is the essence of chaos.

The dripping tap, like the slinky springs earlier, is easy to visualise but extremely difficult to analyse. Chaotic vibrations are found to exist in (analytically) much simpler systems. One such is the driven simple pendulum which is subject to damping. The pendulum is 'driven' by moving the point of suspension back and forth along a horizontal line with s.h.m, and 'damping' is the dissipative effect of friction and air resistance. (These ideas are developed fully in the next chapter.). Such a pendulum may oscillate regularly (non-chaotically) or irregularly (chaotically), depending on the relative values of the dynamical parameters of the system (the frequency and amplitude of the driving force, the degree of damping and the natural frequency of the pendulum itself which is determined by its length). The detailed theory is quite difficult, but it can be shown that the minimum requirement for chaos in a vibrating system is that there should be at least three independent dynamical parameters and that the dynamical equation must contain a non-linear term.

The subject of chaos attracts considerable interest because of its practical implications in (for example) weather forecasting, which has already been alluded to, and for its philosophical implications – touching on such matters as determinism. Sources of further reading are given in the Bibliography.

Problems

1. Show that an alternative solution of the defining equation (2.1) of simple harmonic motion is $x = \alpha \sin \omega t + \beta \cos \omega t$, where α and β are constants. (Note that since there are two arbitrary constants, this is also a general solution.) Comparing this solution with (2.2), express (i) α and β in terms of a and ε and (ii) a and ε in terms of α and β.

Hence, deduce the amplitude of the motion described by $x = 4 \sin \omega t + 3 \cos \omega t$. How could the answer have been obtained by a simple geometric construction? (Note that this is a particular case of the more general problem of adding two s.h.m.s with arbitrary phase difference, which is also accomplished by a simple geometric construction (Chapter 10)).

2. It can be shown that for a simple pendulum (i.e. a heavy mass at the end of a light string fixed at its upper end) of length l, the horizontal displacement x from its equilibrium position is given by $d^2x/dt^2 + (g/l)x = 0$ (so long as $x \ll l$), where g is the acceleration due to gravity.

Show that the period is $2\pi(l/g)^{1/2}$, and deduce the period of a pendulum of length 2 m, taking g to be 10 m s^{-2}. If the bob has mass 0.1 kg, and the amplitude is 0.02 m, deduce (i) the energy of the bob and (ii) the maximum tension in the string.

3. A U-tube having a uniform bore and vertical limbs contains a total length l of liquid. If the liquid in one limb is momentarily depressed, determine the period of subsequent oscillations.

4. A mass is hung from the end of a spring fixed at its upper end, and set into vertical oscillations. It is also allowed to oscillate laterally as a simple pendulum (for which the period is $2\pi(l/g)^{1/2}$). Show that it is impossible for the mass to describe an elliptical path, since this would necessitate that the unstretched length of the spring be zero.

If the frequency of vertical oscillations is double that of horizontal oscillations, show that the equilibrium extension of the spring is one-third of its natural length.

5. Use a calculator to plot the Lissajous figure traced by a particle moving in two dimensions, where $x = 2 \sin \omega t$, $y = \sin 3\omega t/2$. t should start at zero and increase until the path starts to repeat itself. After how many cycles of (i) the x motion and (ii) the y motion does this occur?

If the frequencies of the x and y motions were respectively 5.0 Hz and 5.2 Hz, after how many cycles of each would a repetition of the path now occur? (Do not plot a graph for this part of the question.)

6. Set up an oscilloscope and two sinusoidal oscillators, one connected to the external X input, and the other to the Y input. Keep the frequency of the first fixed at about 100 Hz, and vary that of the second from about 40 Hz to 1 kHz. Observing the screen will give valuable insight concerning the appearance of Lissajous figures.

7. Two stationary railway coaches of masses M and m are coupled together on a level frictionless track by a light coupling of stiffness k. The first is given an impulse at the end remote from the coupling. Find the period of subsequent oscillations.

8. A light spring of stiffness $2k$ is suspended vertically from a rigid support and carries a mass $2m$ at its free end. A second light spring of stiffness k is attached to this mass and carries a mass m at its lower end. Deduce the natural frequencies and the normal modes of oscillation of the system.

    The system is at rest, and the mass $2m$ is suddenly given a vertical velocity u. Derive the equation of motion of this mass.

9. A light elastic string under tension T is stretched between two fixed supports a distance l apart. It carries three particles, each of mass $m/3$, at distances $l/6$, $l/2$ and $5l/6$ from one support. Neglecting the effects of gravity, calculate the frequencies of the normal modes of transverse vibration, indicating their forms.

3 Damped and forced vibrations

3.1 Real vibrating systems

The vibrating systems we considered in Chapter 2 vibrated for ever with undiminished amplitude. Real physical systems do not, of course, vibrate indefinitely. For example the suspended mass on an elastic string of Fig. 2.5 eventually comes to rest; the charge oscillations of the circuit in Fig. 2.6 eventually die away. The equations of motion derived in Chapter 2 (2.10a and 2.12) and the theory leading up to them will now have to be revised to take account of the decay of real vibrations.

The reason for the decay is that the energy originally stored in the system (mechanical potential energy in displacing the suspended mass from its rest position; electrical potential energy in charging the capacitor) is gradually transformed into heat energy by a number of processes. The suspended mass, as it moves, experiences a viscous force because of the air through which it passes. Work has to be done to overcome this force, and hence energy is lost. Even if the mass were in vacuum, there would be hysteresis losses in stretching the elastic string, and also frictional losses where the upper (fixed) end of the string is clamped. In the LC circuit of Fig. 2.6, there will inevitably be some electrical resistance (ignored in Chapter 2) which will convert electrical energy to heat energy by Joule (I^2R) heating.

Before we refine our theory to take account of these effects, we give a physical example of the way the amplitude decreases with time for real systems. Figure 3.1 is from a cathode-ray oscilloscope (CRO) display of charge oscillations for the circuit of Fig. 2.6 when resistance is included. (The plates of C are connected to the Y-plates of a CRO, with the timebase chosen to display a convenient number of oscillations.) Our theory must now be extended fully to account for the features shown.

3.2 Damped oscillations in one dimension

We return to the mass suspended from an elastic string. Consider the situation shown in Fig. 3.2 when the mass is in motion downward, and its instantaneous position is x (measured from the equilibrium level). In addition to the gravitational force mg

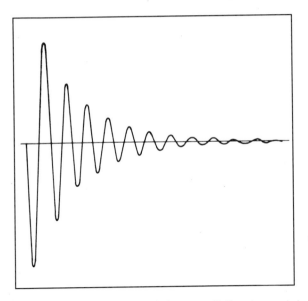

Figure 3.1 Oscilloscope display of charge oscillations in a real circuit.

and the tension $k(l + x)$, we now have a resistive force to take into account. We can at once tell the direction of this force at any instant; since it opposes motion it must be upward in the figure, for the mass is moving downward at the instant shown. What is the magnitude of this resistive force? Here we must appeal to experiment. It is an experimental fact that when a particle moves through a viscous fluid, such as a gas, the resistive force is proportional to the velocity of the particle relative to the fluid, provided the velocity is not too great. Frictional forces can also be shown experimentally to be proportional to relative velocity. The resultant of all the forces which resist the motion may be regarded as a single *damping* force. Further, we will assume that this force at any instant is proportional to the particle velocity v, i.e.

$$\text{Damping force} = Dv = D\frac{\mathrm{d}x}{\mathrm{d}t} \tag{3.1}$$

where D is the coefficient of damping and has the physical dimension $[D] = [MT^{-1}]$. Applying Newton's second law of motion to the particle of Fig. 3.2 gives

$$m\frac{\mathrm{d}^2x}{\mathrm{d}t^2} = mg - k(l + x) - D\frac{\mathrm{d}x}{\mathrm{d}t}. \tag{3.2}$$

Note the signs given to the three forces on the right of this equation: mg is positive because it is downward, that is in the direction of x increasing; the other two are negative because they are upward. Since $mg = kl$ (2.9), (3.2) becomes, on collecting terms,

$$m\frac{\mathrm{d}^2x}{\mathrm{d}t^2} + D\frac{\mathrm{d}x}{\mathrm{d}t} + kx = 0. \tag{3.3}$$

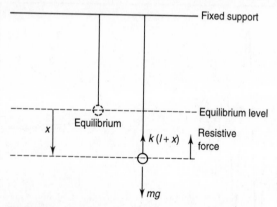

Figure 3.2   Forces on an oscillating suspended mass.

This is the differential equation governing damped vibrations in one dimension (single coordinate x). Before we solve it in order to find the actual equation of motion, it is worthwhile to derive it from energy considerations by an extension of the argument given in section 2.2.1. At the instant shown in Fig. 3.2, the total energy U of the system is the potential energy (PE) plus the kinetic energy (KE), so

$$U = \tfrac{1}{2}m\left(\frac{\mathrm{d}x}{\mathrm{d}t}\right)^2 + \tfrac{1}{2}kx^2. \tag{3.4}$$

In an infinitesimal time $\mathrm{d}t$ during which the mass descends a distance $\mathrm{d}x$, the work $\mathrm{d}W$ done against damping forces is given by force $\times$ distance $= D(\mathrm{d}x/\mathrm{d}t)\,\mathrm{d}x$; therefore

$$\mathrm{d}W = D\left(\frac{\mathrm{d}x}{\mathrm{d}t}\right)\left(\frac{\mathrm{d}x}{\mathrm{d}t}\right)\mathrm{d}t$$

$$= D\left(\frac{\mathrm{d}x}{\mathrm{d}t}\right)^2 \mathrm{d}t. \tag{3.5}$$

This must be equal to the *loss* in energy of the mass and string. Thus

$$\mathrm{d}W = -\frac{\mathrm{d}U}{\mathrm{d}t}\,\mathrm{d}t$$

$$= -\frac{\mathrm{d}}{\mathrm{d}t}\left[\tfrac{1}{2}m\left(\frac{\mathrm{d}x}{\mathrm{d}t}\right)^2 + \tfrac{1}{2}kx^2\right]\mathrm{d}t, \tag{3.6}$$

therefore

$$D\left(\frac{\mathrm{d}x}{\mathrm{d}t}\right)^2 \mathrm{d}t = -\left[m\left(\frac{\mathrm{d}^2x}{\mathrm{d}t^2}\right)\left(\frac{\mathrm{d}x}{\mathrm{d}t}\right) + kx\left(\frac{\mathrm{d}x}{\mathrm{d}t}\right)\right]\mathrm{d}t.$$

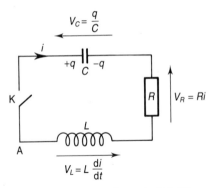

Figure 3.3 *LCR* circuit showing potential differences.

So, cancelling (dx/dt) throughout and collecting terms, we find, as before

$$m\frac{d^2x}{dt^2} + D\frac{dx}{dt} + kx = 0. \tag{3.3}$$

The equation for the electrical circuit of Fig. 3.3 which corresponds to (3.3) is obtained by applying Kirchhoff's second network law to the circuit. The capacitor is initially charged and the key K open. At some instant after closing the key K, suppose the charge in the capacitor is q and the current flowing is i. The potential differences across each element at this instant are as shown. The algebraic sum of potential differences round the circuit going anticlockwise from A to A is zero, so

$$L\frac{di}{dt} + Ri + \frac{q}{C} = 0.$$

Now $i = dq/dt$ and $di/dt = d^2q/dt^2$, so

$$L\frac{d^2q}{dt^2} + R\frac{dq}{dt} + \frac{q}{C} = 0. \tag{3.7}$$

This is the same as (2.11) but with the additional term $Ri = R(dq/dt)$.

Equation (3.7) has exactly the same form as (3.3), so we can point out analogies between the various pairs of corresponding quantities. Displacement x is analogous to instantaneous charge q; velocity $v = (dx/dt)$ to current $i = (dq/dt)$; mass m to inductance L; mechanical damping D to electrical damping, or resistance, R; and mechanical stiffness k to the reciprocal of electrical capacitance $1/C$.

To find exactly how the position x of our mass (or the instantaneous charge q) varies with time, we must find the general solution of (3.3) (or (3.7)). Since both equations have the same form, we can write them both as

$$\boxed{\ddot{x} + 2b\dot{x} + \omega_0^2 x = 0} \tag{3.8}$$

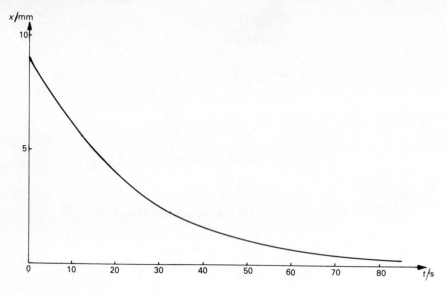

Figure 3.4 Heavy damping. Mass displaced 9 mm and released from rest: $\omega_0 = 0.2\,\mathrm{s}^{-1}$, $b = 0.5\,\mathrm{s}^{-1}$, $A = 9.4$ mm and $B = -0.4$ mm in (3.13).

where

$$2b = D/m \text{ (mechanical case)} \tag{3.9}$$

$$= R/L \text{ (electrical case)} \tag{3.10}$$

$$\omega_0^2 = k/m \text{ (mechanical case)} \tag{3.11}$$

$$= 1/LC \text{ (electrical case)} \tag{3.12}$$

and x in (3.8) represents the displacement or the charge. Note we have used $\dot{x} = dx/dt$, etc., as in Chapter 2. The constants $2b$ and ω_0^2 have been chosen in this way to simplify the subsequent algebra. Although (3.8) has been derived for two specific physical examples, it is of much wider importance in physics. It is a second-order linear differential equation. Its general solution, derived fully in the Appendix, is

$$x = A \exp[-b + \sqrt{(b^2 - \omega_0^2)}]t + B \exp[-b - \sqrt{(b^2 - \omega_0^2)}]t \tag{3.13}$$

where A and B are the arbitrary constants of integration which have to be determined from initial conditions, such as knowledge of the position and velocity of the particle at some instant.

There are three cases to be considered.

Case I: $b > \omega_0$ (heavy damping)
When $b > \omega_0$, we see from (3.9) to (3.12) that, for the vibrating mass $D > 2\sqrt{(mk)}$, or for the *LCR* circuit $R > 2\sqrt{(L/C)}$. In physical terms this means a large amount of

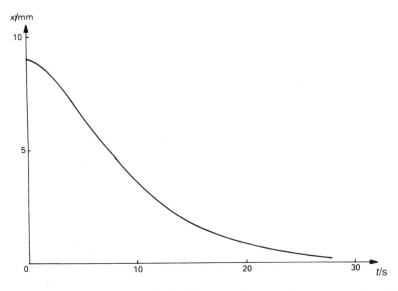

Figure 3.5 Critical damping. Mass displaced 9 mm and released from rest: $\omega_0 = b = 0.2\,\text{s}^{-1}$, $A = 1.8\,\text{mm s}^{-1}$, $B = 9.0\,\text{mm}$ in (3.14).

damping; for example a pendulum suspended in a jar of treacle or a circuit of high resistance. Mathematically, the condition $b > \omega_0$ means that the quantity $b^2 - \omega_0^2$ is positive and the square root is real. Since $-b \pm \sqrt{(b^2 - \omega_0^2)}$ is clearly negative, x is the sum of two exponentially decaying terms. The plot of x against t is typically as shown in Fig. 3.4. There are no oscillations. In the case of the mass on the spring the displaced mass creeps back, or relaxes, to its equilibrium position.

Case II. $b = \omega_0$ (critical damping)
The solution (3.13) breaks down in this case and, as the Appendix shows, the solution is

$$x = (At + B) \exp(-\omega_0 t). \tag{3.14}$$

Again, A and B are the arbitrary constants, determined from initial conditions. There are no oscillations. A typical motion of this kind is shown in Fig. 3.5. It is very similar to that of case I; but, depending on the values of the constants, an overshoot is possible. Motion of this kind is known as 'critically damped'.

EXAMPLE 3.1 Damping in a moving-coil meter
Damping is of great importance in electrical meters for the following reason. Suppose that such a meter initially carrying no current is suddenly connected to a source. The meter will ultimately register a certain value, but it is clearly desirable that it should reach this final reading as quickly as possible. Now this will not happen if the movement is heavily damped so that it responds in a sluggish manner. It is also undesirable that the movement be lightly damped, since the needle will 'overshoot', and perform several

oscillations about the final reading before attaining it. The optimum condition is that the damping should be approximately critical.

The main contribution to the damping in a moving-coil meter is 'electromagnetic damping', and we will assume that other causes of damping (air resistance, etc.) are negligible.

The theory of the moving-coil meter is well known. The coil has N turns, each of area A, and carries a current i. It moves in a radial magnetic field of induction B, and experiences a torque $BANi$. If it has rotated through an angle θ, the suspension exerts a restoring torque $c\theta$, where c is a constant. The equation of motion of the coil is

$$I\frac{\mathrm{d}^2\theta}{\mathrm{d}t} + c\theta = BANi, \tag{3.15}$$

where I is the moment of inertia about the suspension axis.

To calculate the effective damping term, we note that the flux threading the coil can be expressed $BAN\theta$. (The reader may be troubled by this step, but a full justification will not concern us here.) There will be a back e.m.f. equal to the rate of change of this flux, namely $BAN\,\mathrm{d}\theta/\mathrm{d}t$.

If the meter is connected to a source of e.m.f. V, and the total resistance of the circuit is R, then

$$Ri = V - BAN\frac{\mathrm{d}\theta}{\mathrm{d}t}$$

Substitution in (3.15) gives

$$I\frac{\mathrm{d}^2\theta}{\mathrm{d}t^2} + \frac{(BAN)^2}{R}\frac{\mathrm{d}\theta}{\mathrm{d}t} + c\theta = \frac{BAN}{R}V. \tag{3.16}$$

This is the equation (3.8) for damped oscillations. (There should be no cause for concern that the right-hand side is not zero. A simple shift in the origin of θ via the substitution $\theta' = \theta - (BAN/cR)\,V$ would clearly yield an equation in θ' which is identical to (3.16) but with the right-hand side equal to zero.)

By comparison with (3.8), we immediately see that the condition for critical damping ($b = \omega_0$) is

$$\frac{(BAN)^2}{2RI} = \sqrt{\frac{c}{I}}$$

whence the resistance R required for critical damping is

$$R_\mathrm{c} = \tfrac{1}{2}(BAN)^2/(Ic)^{1/2} \qquad\qquad \blacksquare$$

Case III. $b < \omega_0$ *(light damping)*

This is physically the most interesting and important case. The solution is obtained in the Appendix (equation A.15)) and may be written as

$$\boxed{\begin{aligned} x &= a\,\mathrm{e}^{-bt}\sin(\omega t + \varepsilon) \\ \text{where } \omega &= \sqrt{(\omega_0^2 - b^2)}. \end{aligned}} \tag{3.17}$$

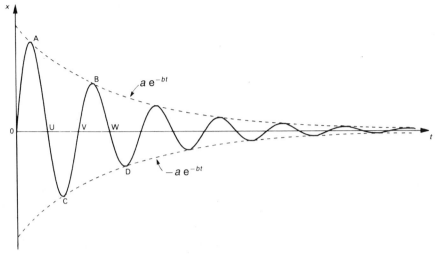

Figure 3.6 Damped s.h.m.: plot of (3.17).

Now (3.17) has the same form as (2.2), but the amplitude term a in (2.2) has been replaced by $a\,e^{-bt}$, and the circular frequency term has become $\sqrt{(\omega_0^2 - b^2)}$.

Let us look at these in turn. We see that the amplitude of the motion decreases exponentially with time on account of the e^{-bt} term. Further, we see that there are now oscillations on account of the $\sin(\omega t + \varepsilon)$ term. A plot of (3.17) is shown in Fig. 3.6. This has the same shape as Fig. 3.1, so our theory seems to be producing sensible answers. What we have is something very similar to s.h.m., but with the amplitude getting less and less, and after an infinite time disappearing altogether. Mathematically, we have the product of two terms $a\,e^{-bt}$ and $\sin(\omega t + \varepsilon)$ at each value of t; the curve is said to lie between the asymptotes $\pm a\,e^{-bt}$.

What is the period of this motion? Here we have a difficulty, for the motion never exactly repeats itself. The point A (Fig. 3.6) is never again reached. However, by comparison with undamped s.h.m. we can define the period as the elapsed time between a zero of x and the next-but-one zero of x, such as the points U and W in Fig. 3.6. Clearly

$$\omega T = 2\pi, \text{ or } T = 2\pi/\omega$$

Thus the period is

$$T = \frac{2\pi}{\omega} = \frac{2\pi}{\sqrt{(\omega_0^2 - b^2)}}. \tag{3.18}$$

If the damping is very light, so that $\omega_0^2 \gg b^2$, we see that

$$T = \frac{2\pi}{\omega} \simeq \frac{2\pi}{\omega_0}. \tag{3.19}$$

We can find the values of x at the various maxima and minima by the usual process of equating $\dot{x} = dx/dt$ to zero. Thus from (3.17) we see

$$\frac{dx}{dt} = -ba\,e^{-bt}\sin(\omega t + \varepsilon) + \omega a\,e^{-bt}\cos(\omega t + \varepsilon)$$

$$= a\,e^{-bt}[-b\sin(\omega t + \varepsilon) + \omega\cos(\omega t + \varepsilon)]$$

$$= 0 \text{ at maxima and minima,}$$

so

$$\tan(\omega t + \varepsilon) = \omega/b. \tag{3.20}$$

Suppose (3.20) is satisfied at a particular time when x is maximum. Then since ω and b are constants, the succeeding times at which it is again satisfied are those for which $\tan(\omega t + \varepsilon)$ takes the same value. This will clearly be when ωt has increased by π, 2π, 3π, etc., i.e. when t has increased by π/ω, $2\pi/\omega$, $3\pi/\omega$, etc. Since maxima and minima alternate (see Fig. 3.6), it is clear that π/ω, $3\pi/\omega$, $5\pi/\omega$, etc. correspond to minima, and $2\pi/\omega$, $4\pi/\omega$, $6\pi/\omega$, etc. correspond to maxima. The period T is, of course, $2\pi/\omega$. Consider two successive maxima. Since $(\omega t + \varepsilon)$ has increased by 2π, $\sin(\omega t + \varepsilon)$ must be the same for both. Thus

$$\frac{x_n}{x_{n+2}} = \frac{a\,e^{-bt}}{a\,e^{-b(t+T)}} = e^{bT}. \tag{3.21}$$

The right-hand side of (3.21) is independent of n, which tells us this is the ratio of any pair of successive maxima, or successive minima. (It should now be obvious that the ratio of a maximum to the following minimum is $-\exp(\pi b/\omega)$.) The successive maxima thus form a geometrical progression of constant ratio e^{bT}, i.e.

$$\frac{x_1}{x_3} = \frac{x_3}{x_5} = \ldots = \frac{x_n}{x_{n+2}} = e^{bT}. \tag{3.22}$$

There are several parameters which may be chosen to describe the manner in which a damped vibrating system decays. Prominent among them are the logarithmic decrement λ, the time-constant τ and the quality factor Q.

Logarithmic decrement is defined as

$$\boxed{\lambda = \ln\!\left(\frac{x_n}{x_{n+2}}\right),} \tag{3.23}$$

that is, as the natural logarithm of the ratio of successive maxima. Thus, from (3.21)

$$\lambda = bT. \tag{3.24}$$

The time-constant τ is the time taken for the amplitude to fall to $1/e$ of its initial value. It follows from (3.22) that

$$\tau = 1/b. \tag{3.25}$$

The quality factor Q is defined as $2\pi \div$ (the fraction of the energy of the system lost to damping forces in one cycle). Thus

$$Q = 2\pi \times \frac{\text{energy in system at start of cycle}}{\text{energy lost during cycle}}. \tag{3.26}$$

If our system is the suspended vibrating mass, then at the instant the mass is at the nth (maximum) displacement x_n, its velocity is zero and the energy of the system is entirely potential and of value $\frac{1}{2}kx_n^2$ where k is the stiffness of the string. The energy lost in one cycle is thus $\frac{1}{2}k(x_n^2 - x_{n+2}^2)$, and

$$Q = 2\pi \times \frac{\frac{1}{2}kx_n^2}{\frac{1}{2}k(x_n^2 - x_{n+2}^2)}$$

$$= 2\pi \left/ \left(1 - \frac{x_{n+2}^2}{x_n^2}\right) \right.$$

$$= \frac{2\pi}{(1 - e^{-2bT})} \text{ by (3.21).} \tag{3.27}$$

If the damping is light, so that b is small,

$$e^{-2bT} \simeq 1 - 2bT,$$

so

$$Q = \frac{2\pi}{2bT} = \frac{\pi}{bT}. \tag{3.28}$$

Also, from (3.25)

$$Q = \pi\tau/T = \pi \times \text{(number of oscillations in one time-constant).} \tag{3.28a}$$

Q is thus independent of the actual energy possessed by the vibrating system at any given instant. The quality factor Q has greater importance in forced damped vibrations, but it is worth introducing the idea at this stage.

3.3 Forced vibrations

The preceding sections of this chapter treat vibrating systems which, once set into vibration (for example by displacing a mass or charging a capacitor), the system is left to itself. There is no further interference by the outside world; in particular no further energy is introduced. Such vibrations are sometimes referred to as free vibrations, and the frequency of such vibrations is entirely determined by the system itself (see, for example, (3.18) above).

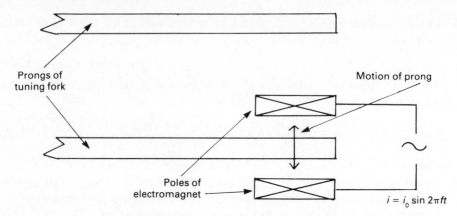

Figure 3.7   Forced vibration of a tuning fork.

When a vibrating system is subjected to a continuous periodic disturbance of some kind, the resulting vibrations are quite different. For example, when a tuning fork is arranged as shown in Fig. 3.7 with a prong between the poles of an electromagnet fed with alternating current (a.c.), the fork sounds not with its own natural frequency but with the frequency f of the alternating current. Forced vibrations of this kind are of immense importance throughout physics and engineering, and we shall treat them in some detail.

We shall expand our simple mechanical and electrical systems of Figs 3.2 and 3.3 to incorporate driving forces as shown in Fig. 3.8. In the mechanical system, a sinusoidally varying vertical force is applied to the now-familiar mass m suspended from an elastic string of stiffness k. The force has maximum value F_0 and varies sinusoidally with time with circular frequency p. A possible practical method of applying a force of this kind is an a.c.-fed electromagnet (as in Fig. 3.7). The electrical case is more straightforward. The original circuit of Fig. 3.3 is simply driven by an a.c. source, having zero impedance, of peak e.m.f. E_0 and circular frequency p as shown in Fig. 3.8b. When we include the driving force in (3.3) we obtain, by a straightforward application of Newton's second law of motion,

$$m\ddot{x} + D\dot{x} + kx = F_0 \sin pt. \tag{3.29}$$

When we apply Kirchhoff's second network law (the e.m.f. in a circuit is equal to the algebraic sum of the potential differences round the circuit at any instant) we obtain

$$L\ddot{q} + R\dot{q} + (1/C)q = E_0 \sin pt. \tag{3.30}$$

The equations (3.29) and (3.30) have exactly the same mathematical form, and we write for both of them

$$\boxed{\ddot{x} + 2b\dot{x} + \omega_0^2 x = P \sin pt} \tag{3.31}$$

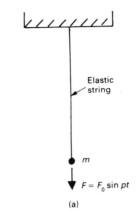

(a)

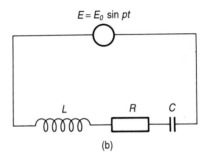

(b)

Figure 3.8 Driven systems: (a) mechanical and (b) electrical.

where $2b$ and ω_0^2 are still given by (3.9) to (3.12) above, and now

$$P = F_0/m \text{ (mechanical case)} \tag{3.32}$$

$$P = E_0/L \text{ (electrical case)}. \tag{3.33}$$

The second-order, linear differential equation (3.31) is solved towards the end of the Appendix where it is shown that the general solution (for light damping) is

$$x = a\,e^{-bt}\,\sin(\omega t + \varepsilon) + \frac{P\,\sin(pt - \delta)}{[(\omega_0^2 - p^2)^2 + 4p^2b^2]^{1/2}} \tag{3.34}$$

where

$$\delta = \tan^{-1}\left(\frac{2bp}{\omega_0^2 - p^2}\right). \tag{3.35}$$

Here a and ε are arbitrary constants, and we define $\omega^2 = \omega_0^2 - b^2$ as we did in (3.17). Now (3.34) consists of two terms. The first term

$$a\,e^{-bt}\,\sin(\omega t + \varepsilon) \tag{3.35a}$$

is the solution we have already obtained (3.17) when there is no driving force. This is known as the complementary function (CF). Note that the circular frequency ω of this term is that of the free vibrations of the system, and note also the presence of the damping term e^{-bt}. The form of the CF is shown in Fig. 3.6. Note too that the arbitrary constants a and ε appear entirely in this term; it is therefore this *damped* term which depends on the initial conditions. The second term

$$\frac{P\,\sin(pt - \delta)}{[(\omega_0^2 - p^2)^2 + 4p^2b^2]^{1/2}} \tag{3.35b}$$

is known as the particular integral (PI) of (3.31). It has the circular frequency p of the driving force, and a constant (undamped) amplitude

$$\frac{P}{[(\omega_0^2 - p^2)^2 + 4p^2b^2]^{1/2}}.$$

The PI is completely independent of the arbitrary constants a and ε, and therefore of the initial conditions. Note that the PI is of the same frequency as the driving force but differs in phase from the driving force by the (constant) phase angle δ (3.35).

The CF, PI and their sum (the full general solution, and hence the actual equation of motion) are shown in Fig. 3.9. The net displacement (Fig. 3.9c) at any instant is given by simply adding the ordinates in Fig. 3.9a and Fig. 3.9b. In the early stages of the motion the resultant is a form of 'beats' (see the end of section 2.3.1), the usual consequence of compounding two vibrations of unequal frequency. The contribution of the CF (Fig. 3.9a) diminishes with time until, after a time of t_A (where the first curve has almost vanished) from the start of the motion, this term has effectively disappeared altogether as a result of damping. After t_A, we are left with the steady PI term only. The motion thus comprises two distinct stages: (a) the *transient* stage when the motion is the beat resultant of the CF and PI, and (b) the *steady-state* stage where the motion is a steady s.h.m. of amplitude A given by

$$A = \frac{P}{[(\omega_0^2 - p^2)^2 + 4p^2b^2]^{1/2}} \tag{3.36}$$

and of circular frequency p equal to that of the driving force. Note that both A and p are independent of initial conditions. This is known as forced s.h.m.

The time for which the transients persist is determined by b, and hence by the damping factor. The greater the value of b, the more quickly do the transients die away. (The transient amplitude falls to e^{-1} of its initial value in a time $1/b$.) As

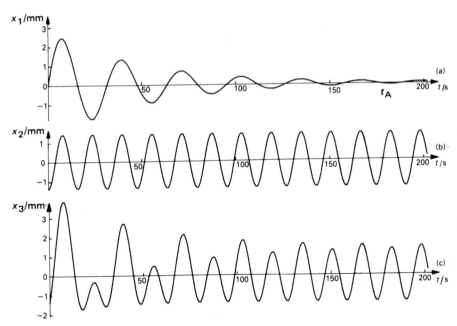

Figure 3.9 Forced damped vibrations. (a) C.F.(3.35a) with $a = 3.0$ mm, $\omega_0 = 0.2$ s^{-1}, $b = 0.02$ s^{-1} and $\varepsilon = 0$. (b) PI (3.35b) with $p = 0.4$ s^{-1} and amplitude $P/[(\omega_o^2 - p^2)^2 + 4p^2b^2]^{1/2} = 1.5$ mm. (c) General solution (i.e. sum of CF and PI with $x_3 = x_1 + x_2$ at each value of t).

emphasised earlier, the transients depend upon the initial conditions, because they involve the constants a and ε which require knowledge of the initial conditions for their determination. The relative importance of transients depends on the nature of the vibrating system. In a pipe organ (driven vibrations of a column of air in a pipe) the starting transients are of crucial importance and determine the character of the perceived sound; if a tape of organ music is played backwards, the sound is totally different. For a.c. circuits, on the other hand, the starting transients can usually be ignored, and traditional treatments of alternating current are concerned only with the steady forced vibrations of charge in the circuit.

3.4 Resonance

3.4.1 Displacement and charge resonance

We now examine how the amplitude of steady-state forced vibrations varies with the circular frequency p of the applied force. As we saw (3.34), in the steady-state

$$x = A \sin(pt - \delta) \tag{3.37}$$

where

$$A = \frac{P}{[(\omega_0^2 - p^2)^2 + 4p^2b^2]^{1/2}}. \tag{3.38}$$

How does the amplitude A vary with p? And, in particular, for what value of p is A maximum, and what is this maximum value? Now A is maximum when

$$y = (\omega_0^2 - p^2)^2 + 4p^2b^2 \tag{3.39}$$

is minimum.

The condition for a stationary value, maximum or minimum, is $dy/dp = 0$. Thus

$$dy/dp = -4p(\omega_0^2 - p^2) + 8pb^2 = 0$$

or

$$p^2 = \omega_0^2 - 2b^2 \tag{3.40}$$

which will be satisfied so long as $\omega_0^2 \geqslant 2b^2$. If this were not the case, there would be no maximum or minimum, but this is a situation of no physical interest which will not be considered further. We could, if we wish, check that condition (3.40) is for maximum A (minimum y) by differentiating again and testing the sign of d^2y/dp^2.

Thus the maximum amplitude is obtained when the frequency of the applied force is

$$\boxed{f = \frac{p}{2\pi} = \frac{\sqrt{(\omega_0^2 - 2b^2)}}{2\pi}.} \tag{3.41}$$

This value lies between the frequency for free undamped vibrations, $\omega_0/2\pi$, and that for free damped vibrations $\sqrt{(\omega_0^2 - b^2)}/2\pi$ (3.18).

We obtain the value for maximum amplitude A_0 by substituting $p = \sqrt{(\omega_0^2 - 2b^2)}$ from (3.40) into (3.38). Thus

$$A_0 = \frac{P}{2b(\omega_0^2 - b^2)^{1/2}}. \tag{3.42}$$

This phenomenon of the amplitude taking a maximum value is known as resonance. We say that the response of the vibrating system to the applied force (as demonstrated by the maximum amplitude of vibration produced) is greatest at the resonant circular frequency given by (3.40). A typical variation of amplitude A of forced s.h.m. with applied circular frequency is shown in Fig. 3.10. In physical terms this represents the variation in maximum displacement from equilibrium of the driven suspended mass (Fig. 3.8a) or the maximum instantaneous value of charge on the capacitor in the circuit of Fig. 3.8b. Note (a) the resonant frequency is not the same as either the free or the damped natural frequency of the vibrating system, (b) the resonant frequency decreases as the damping increases, and (c) for light damping (small b) the resonance curve is sharp and high, while for heavy damping (large b), the curve is broad and low.

The type of resonance encountered here is referred to as displacement resonance (in the mechanical case) or charge resonance (in the electrical case) to distinguish

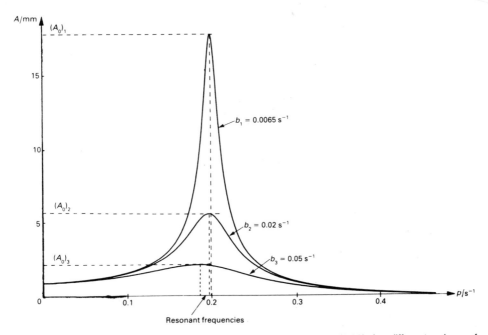

Figure 3.10 Amplitude resonance curves. Variation of A with p (3.38) for different values of b; $\omega_0 = 0.2 \text{ s}^{-1}$, $P = 0.047 \text{ mm s}^{-2}$. Note that the resonant frequency decreases as b increases.

them from resonance of another character which we treat in the following section. The variation of the phase angle δ (3.37) between the driving force and the forced motion is of interest. As p is increased steadily from zero, the variation of δ with p, given by (3.35), is as shown in Fig. 3.11.

When $p \ll \omega_0$ the driving force and the motion are in phase ($\delta = 0$). As p increases, the motion increasingly lags in phase. At $p = \omega_0$ the phase lag is $90°$ and we say that the driving force and the motion are in quadrature. Finally, when $p \gg \omega_0$, the motion lags in phase by $180°$ and is in antiphase with the driving force. The variation of δ with p is gradual for heavy damping (large b), but becomes increasingly abrupt as the damping decreases.

3.4.2 Velocity and current resonance

As the frequency of the applied force in the mechanical system of Fig. 3.8a is varied, then as well as the amplitude of the resulting forced vibrations changing in the manner we have just seen, the maximum *velocity* of the particle will change also. This too has a maximum, and we speak of *velocity resonance*. Similarly the current in the circuit of Fig. 3.8b will have a maximum as we vary p, and this is known as current resonance. It might be thought that velocity (current) will resonate at the same

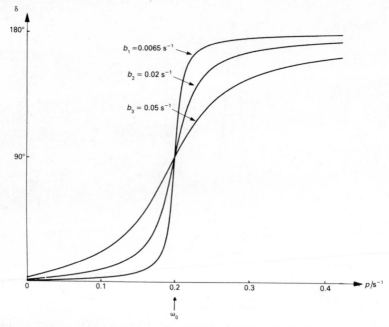

Figure 3.11 Variation of phase angle δ with applied circular frequency (3.35) for different values of b: $\omega_0 = 0.2\,\mathrm{s}^{-1}$.

frequency as the amplitude (charge), but, as we shall show, this turns out not to be the case. We can treat velocity and current together, for they are analogous, i.e.

$$\text{velocity} = \mathrm{d}x/\mathrm{d}t \tag{3.43}$$

and

$$\text{current } i = \mathrm{d}q/\mathrm{d}t. \tag{3.44}$$

We obtain an expression for $\mathrm{d}x/\mathrm{d}t$ directly from (3.34). Let us assume the steady state has been reached so we can ignore the first term in (3.34). Thus

$$v = \frac{\mathrm{d}x}{\mathrm{d}t} = \frac{Pp\,\cos(pt - \delta)}{[(\omega_0^2 - p^2)^2 + 4p^2b^2]^{1/2}} \tag{3.45}$$

or

$$v = v_0\,\cos(pt - \delta), \tag{3.46}$$

where

$$v_0 = P \Big/ \left[\left(\frac{\omega_0^2 - p^2}{p}\right)^2 + 4b^2\right]^{1/2}. \tag{3.46a}$$

Note that we have divided top and bottom by p.

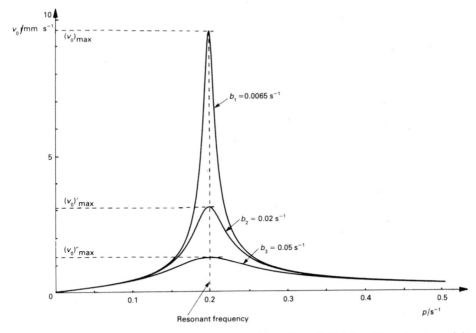

Figure 3.12 Velocity resonance curves. Variation of v_0 with p (3.46a) for different values of b: $\omega_0 = 0.2\,\mathrm{s}^{-1}$ and $P = 0.125\,\mathrm{mm\,s}^{-2}$. Note the same resonant frequency for each value of b.

The condition for maximum value of v_0 can be seen very easily; it is $p = \omega_0$, the undamped circular frequency of the system. The maximum value of the velocity $(v_0)_{max}$ is given by

$$(v_0)_{max} = \frac{P}{2b}. \tag{3.47}$$

We note that the resonant frequency is independent of damping. The resonance curves are shown in Fig. 3.12. As with amplitude resonance, light damping produces a sharp, high resonance, while heavy damping produces a low, broad resonance.

3.5 Power to maintain forced vibrations

We can most conveniently find the energy expended in maintaining steady-state forced vibrations by considering the suspended mass of Fig. 3.8a to which is applied a sinusoidally varying force $F = F_0 \sin pt$.

In a displacement $\mathrm{d}x$, the driving source expends an amount of work

$$\mathrm{d}W = F_0 \sin(pt)\,\mathrm{d}x.$$

In one complete cycle, the amount of work done is

$$W = \int_{1 \text{ cycle}} F_0 \sin(pt) \, dx = \int_0^{2\pi/p} F_0 \sin(pt) \frac{dx}{dt} \, dt$$

since the period T is equal to $2\pi/p$.

Substituting dx/dt from (3.45) and using (3.32) we have

$$W = \frac{pF_0^2}{m[(\omega_0^2 - p^2)^2 + 4p^2 b^2]^{1/2}} \int_0^{2\pi/p} \sin(pt) \cos(pt - \delta) \, dt$$

If we expand $\cos(pt - \delta)$, note that

$$\int_0^{2\pi/p} \sin^2(pt) \, dt = \pi/p,$$

$$\int_0^{2\pi/p} \sin(pt) \cos(pt) \, dt = 0$$

and

$$\sin \delta = \frac{2bp}{[(\omega_0^2 - p^2)^2 + 4p^2 b^2]^{1/2}}$$

(from 3.35); we arrive eventually at

$$W = \frac{2\pi F_0^2 pb}{m[(\omega_0^2 - p^2)^2 + 4p^2 b^2]} \tag{3.48}$$

The equation (3.48) represents the net expenditure of energy by the driving source in one cycle. This must therefore be the work done in overcoming damping forces in one cycle. We can find the mean power consumption as follows: (3.48) gives the total energy consumed in one complete cycle of duration $T = 2\pi/p$. The mean power $\mathscr{P}$ is therefore given by

$$\mathscr{P} = \frac{W}{T} = \frac{p}{2\pi} \frac{2\pi F_0^2 pb}{m[(\omega_0^2 - p^2)^2 + 4p^2 b^2]},$$

or

$$\mathscr{P} = \frac{F_0^2 p^2 b}{m[(\omega_0^2 - p^2)^2 + 4p^2 b^2]}. \tag{3.49}$$

If (3.49) is written in the form

$$\mathscr{P} = \frac{F_0^2 b/m}{\dfrac{(\omega_0^2 - p^2)^2}{p^2} + 4b^2} \tag{3.50}$$

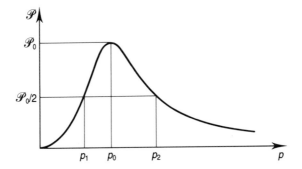

Figure 3.13 Power resonance curve.

we see that the power $\mathscr{P}$ has its maximum value (as p varies) when the first term in the denominator of (3.50) is zero; that is, when $p = \omega_0$. This is exactly the same condition as for velocity resonance, as was shown above in the discussion immediately preceding (3.47). Thus power resonance (maximum energy transfer) and velocity resonance occur at the same (driver) frequency. The maximum value for the power, $\mathscr{P}_0$, is obtained by putting $\omega_0 = p$ in (3.50). Thus

$$\mathscr{P}_0 = \frac{F_0^2}{4mb}. \tag{3.51}$$

A typical power resonance curve is shown in Fig. 3.13 in which p_0 $(=\omega_0)$ is the resonant circular frequency, and p_1 and p_2 are the circular frequencies at which the power $\mathscr{P}$ is one-half of the resonant value. A convenient measure for the sharpness of the resonance is the ratio

$$p_0/(p_2 - p_1). \tag{3.52}$$

Now $\mathscr{P}$ falls to half its maximum value when the two terms in the denominator of (3.50) are equal. Thus

$$(\omega_0^2 - p^2)^2 = 4p^2b^2, \tag{3.53}$$

or

$$\omega_0^2 - p^2 = \pm 2pb$$

and

$$(\omega_0 + p)(\omega_0 - p) = \pm 2pb. \tag{3.54}$$

If the resonance is fairly sharp we may make the approximations $p \approx p_0$, $\omega_0 + p \approx 2p_0$, $\omega_0 - p \approx (p_2 - p_1)/2$. Thus

$$\frac{p_0}{p_2 - p_1} = \frac{p_0}{2b} = \frac{\omega_0}{2b}. \tag{3.55}$$

But the quality factor Q (section 3.2 above) was found to be

$$Q = \pi/bT. \qquad [3.28]$$

So, since $T = 2\pi/\omega_0$, we see $Q = p_0/(p_2 - p_1)$ or, in terms of actual frequencies ($p_0 = 2\pi f_0$, etc.), we have

$$\boxed{Q = \frac{f_0}{f_2 - f_1}} \qquad (3.56)$$

The quantity $(f_2 - f_1)$ is known as the *bandwidth* of the system. The Q factor has a useful interpretation in terms of amplification. In the case of displacement resonance, we found that the amplitude A_0 at resonance is given by

$$A_0 = \frac{P}{2b(\omega_0^2 - b^2)^{1/2}}. \qquad [3.42]$$

If we write $P = F_0/m$, and assume light damping ($\omega_0^2 \gg b^2$), this becomes

$$A_0 = \frac{F_0}{2bm\omega_0}. \qquad (3.57)$$

If we substitute for b from (3.28) we obtain

$$A_0 = \frac{QF_0}{m\omega_0^2}, \qquad (3.58)$$

or

$$A_0 = \frac{QF_0}{k} \qquad (3.59)$$

where k, the stiffness, is substituted from (3.11). Now k is the restoring force for unit displacement. A slowly applied force F_0 would produce an equilibrium displacement F_0/k; (3.59) shows us that at resonance we have a maximum displacement of Q times this value. Hence Q can be regarded as a displacement amplification factor.

Q has perhaps its greatest importance in relation to electrical circuits. If we rewrite the above relationships in electrical quantities it is easy to show that, for the circuit of Fig. 3.8b:

$$Q = \frac{\omega_0 L}{R} \qquad (3.60)$$

and

$$\omega_0 = \frac{1}{\sqrt{(LC)}}. \qquad (3.61)$$

The Q of a circuit determines its ability to select a narrow band of frequencies from a wide range of input frequencies. This is obviously important for radio receivers. The 'selectivity' of the tuner stage in a receiver is its ability to select the required signal only, and this ability is, in turn, determined by the Q for the circuit.

Radio receivers operating in the MHz region have Q values of several hundreds. Microwave cavities have Q values of the order of 10^5.

EXAMPLE 3.2 How much power is expended by a child on a swing?

We can perform an order of magnitude calculation of this power using some of the ideas which have been developed. The swing is a damped oscillator which is maintained at constant amplitude by a periodic force exerted by the child. The mechanism whereby energy is transferred from the child to the swing is of some interest. A simplified model of the action is that at the lowest point of the motion, the child throws forward his or her legs and pulls on the supporting chains or ropes. This reduces the distance from the centre of mass to the support bar, and therefore reduces the moment of inertia I. Suppose this change takes place instantaneously. Then in that instant the angular momentum $I\omega$ is unchanged, so the angular velocity ω increases. The kinetic energy $\frac{1}{2}I\omega^2 = \frac{1}{2}(I\omega)\omega$ therefore increases. The legs and chains are relaxed to their original positions when the swing is at its highest point.

An alternative, but perhaps less respectable way of looking at the action is to realise that at the lowest point when the legs are being thrown forward, work is being done by the child against the 'centrifugal force', so energy is being introduced into the system.

Now the distance l from the centre of mass of the child to the point of support is approximately 2.5 m, and the periodic time is about 3 s. If the child were to stop applying the force, the oscillations would die away with a time constant of (typically) about 80 s. Hence, Q for this oscillator, which is π times the number of oscillations in this time, is about 80. But

$$Q = \frac{2\pi \times \text{stored energy}}{\text{energy dissipated per cycle}} \text{ from (3.26).}$$

Here, the stored energy is $mgl(1 - \cos\theta)$, where m is the mass of the child, and θ the amplitude of swing. Taking $m \approx 40$ kg, $g \approx 10$ m s^{-2} and $\theta \approx 45°$, we find that the stored energy is about 300 J. Hence,

$$\text{energy dissipated per cycle} \approx \frac{2\pi \times 300}{Q} \approx 25 \text{ J}$$

Finally, since the period ≈ 3 s, the mean power expended in maintaining the oscillations is ~ 8 W. ∎

EXAMPLE 3.3 Emission and absorption of light by atoms

The study of light emitted by a source is very important in physics since it gives valuable insight into the nature of the source, and the physical processes concerned with the emission. A hot solid (for example, a tungsten filament in a lamp) emits radiation predominantly in visible and infrared regions. Because of the strong forces between the atoms in a solid, the radiation cannot be regarded as arising from free atoms, but rather from an assembly of strongly coupled atoms.

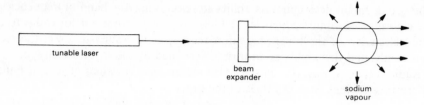

Figure 3.14 Resonance fluorescence.

If, however, the light source is a suitably excited atomic vapour of low density, the emission is from an assembly of free atoms, and the observed radiation is directly related to the nature of a single atom. A familiar example of such a source is a sodium lamp used for street lighting. Such a source emits radiation at several sharply defined wavelengths.

The emission of light by an atom is a process which is intimately bound up with the quantum theory, and usually developed from this standpoint. Nevertheless, it is possible to understand some aspects of the subject by making the very simple postulate that *a free atom behaves like a classical system of particles capable of oscillating in several normal modes*. If the atom is excited so as to vibrate in one of these normal modes, it emits radiation of the same frequency.

In the case of sodium, the predominant wavelengths are 589.0 and 589.6 nm, and the corresponding frequencies c/λ are 5.090×10^{14} and 5.085×10^{14} Hz. (There are several other wavelengths emitted, but we ignore these for simplicity.)

Now since the atom is emitting radiation, it loses energy, and the motion of the oscillator is exponentially damped. In the case of sodium, the amplitude decays with a time-constant $\tau \sim 3 \times 10^{-8}$ s. The Q value is therefore (from (3.28a))

$$\pi \times 5 \times 10^{14} \times 3 \times 10^{-8} \approx 5 \times 10^{7}.$$

If an atom can be thought of as a damped oscillator, can it exhibit forced vibrations? The answer is that it can, but since Q is so high, the amplitude of forced vibrations is very small unless the driving frequency is very close to the natural frequency. This is illustrated by considering sodium vapour of low density, illuminated by light from a tunable laser (Fig. 3.14).

Unless the laser frequency is very close to one of the resonant frequencies of the sodium atom, the light passes through the vapour with very little attenuation. But if the laser is carefully tuned to a resonant frequency, the sodium atoms absorb the light, and subsequently re-emit it in all directions. This process is known as *resonance fluorescence*. From (3.56) the accuracy of tuning must be better than 1 part in 10^6.

It should be stated that our analysis is rather over-simplified, since it does not consider the Doppler shift in resonant frequency of the atom due to its motion.

We emphasise too that this treatment of resonance fluorescence, while appropriate to the scope of this book, is no substitute for the fuller and more satisfactory analysis to be found in texts on spectroscopy. ∎

3.6 Piezoelectric vibrations in quartz

It is found that for certain crystals of low symmetry (more specifically for those which lack a centre of symmetry), the application of a longitudinal stress produces electric polarisation. This is the *piezoelectric effect* and the best known piezoelectric substances are quartz and barium titanate. The phenomenon is capable of quite simple explanation at the atomic level. Because of the absence of symmetry, when the specimen is subjected to pressure, the redistribution of the electron cloud consequent upon the shortening of bonds causes a net flow of negative charge in one direction, resulting in a layer of positive charge at one end of the specimen and a layer of negative charge at the other – a process known as polarisation.

Expressed in simplest terms, the effect means that a potential difference can be produced across a slab of quartz simply by squeezing it. This is made use of in, for example, piezoelectric gas lighters, in which the potential difference is sufficient to produce a spark, and gramophone pickups, in which the variations in depth of the record groove are transformed into voltage variations which are amplified to produce sound. Of more interest in the context of vibrations is the converse piezoelectric effect in which the application of an electric field to a piezoelectric crystal causes a strain or change in dimensions. The applied field E is proportional to the resulting strain ε so we can write

$$\varepsilon = dE \tag{3.62}$$

where d is the piezoelectric modulus and has the unit m V^{-1}. This follows at once from (3.62) since ε is a dimensionless ratio (the fractional change in dimension) and E has the unit V m^{-1}. The value of d for quartz is 2.3×10^{-12} m V^{-1}. Thus if a potential difference of 100 V is applied across a slab of quartz of thickness 5 mm, producing an electric field of intensity 2×10^4 V m^{-1}, the resulting strain, by (3.62), is

$$2.3 \times 10^{-12} \times 2 \times 10^4 = 4.6 \times 10^{-8},$$

and the actual change in dimension of the slab is 2.3×10^{-10} m.

Consequently, quartz crystals are used extensively as *transducers* for producing minute and readily controllable movements by means of voltage changes which are easy to make. In particular, piezoelectric transducers are used extensively for production of *ultrasonic waves*, which are acoustical waves of very high frequency, and frequencies of up to 10 GHz can be achieved.

The converse piezoelectric effect is the basis of quartz oscillators which provide high-precision frequency references in the following way. A piece of quartz may be cut so that its resonant frequency for mechanical vibrations is of a required value. When such a specimen is placed between the plates of a capacitor, a very sharp resonance will occur when the frequency of the applied potential difference (p.d.) matches that of the mechanical resonance frequency of the specimen. This enables the electrical frequency to be tuned very precisely.

This brief account of the piezoelectric effect is much simplified. Piezoelectric crystals are *anisotropic*, i.e. they have different properties in different directions, and piezoelectric behaviour cannot be described in terms of a single modulus as we have done. A full treatment requires the use of *tensors* which are mathematical devices for treating direction-dependent properties. Our discussion of quartz, though much oversimplified, is in fact valid for a crystal cut in a special way.

Problems

1. A simple pendulum of length l has a spherical bob of radius r and density ρ, which is immersed in a bath of liquid of density ρ_0 and viscosity η. Show that for the oscillations to be critically damped,

$$r = \frac{3}{2}\left[\frac{\eta^2 l}{\rho(\rho - \rho_0)g}\right]^{1/4}.$$

(The damping force is given by Stokes's formula $6\pi\eta rv$, where v is the velocity.)

If r is greater than this value, will the motion be more, or less, than critically damped?

Calculate the value of r for critical damping in water, if $l = 0.5$ m and $\rho = 5 \times 10^3$ kg m^{-3}. Take $\eta = 10^{-3}$ kg m^{-1} s^{-1} and $g = 10$ m s^{-2}.

2. Show that the time-constant for the decay of energy of an exponentially damped oscillator is one half of the time-constant for the decay of amplitude.

3. A critically damped oscillator is released from rest at time $t = 0$ at a distance a from its equilibrium position. Show that its equation of motion is $x = a(1 + bt)e^{-bt}$, where b is the constant used in this chapter.

Prove that the oscillator does not 'overshoot' the equilibrium position.

4. A critically damped oscillator is released at time $t = 0$ at a distance a (>0) from its equilibrium position, with velocity v_0 towards this position. Show that 'overshoot' will occur if $v_0 > ab$.

Taking $a = 1$, $b = 2$ and $v_0 = 4$, plot a graph of the displacement against time.

5. A simple pendulum of length 1 m is set into oscillation with amplitude 0.05 m. After 5 min, the amplitude has fallen to 0.025 m. Deduce (i) the constant b in the nomenclature of this chapter, (ii) the time-constant, (iii) the logarithmic decrement, (iv) the Q value, (v) the energy dissipated per cycle when the amplitude was 0.05 m, if the mass of the bob is 0.1 kg.

If an external device were used to sustain the oscillations, to approximately what accuracy should the frequency of this device be matched to the natural frequency of the pendulum?

6. Perform the following experiment. Take a length of string about 50 cm, and tie a mass of about 20 g to one end. Hold the other end steady, and observe the oscillations. Now move the upper end of the string to and fro at a significantly higher frequency. After a little time, the mass will be moving at this latter frequency, approximately in antiphase with the movement of the hand.

Now reduce the driving frequency, and observe the resonance which occurs when it is near the natural frequency. The phase difference is then about $90°$, the driving force leading the mass.

It is possible, but less easy, to show that when the driving frequency is significantly below the natural frequency, the motions of the mass and the hand are approximately in phase.

7. The damper of a particular piano string may be held off by depressing the corresponding key, or by operating the sustaining pedal. When this is done, if a clarinet near the string emits a quick burst of the same note, the string will produce a faint sound. Explain this.

   Estimate the Q value of the string by a quick experiment. Hence estimate the closeness in frequency to which the clarinet and piano must be tuned to achieve the above behaviour.

8. An inductance 0.1 H, a capacitor 10^{-7} F, and a resistor R are all connected in series. For what range of values of R will no oscillations occur?

   If $R = 100\ \Omega$, deduce (i) the frequency of the damped oscillations, (ii) the time-constant, (iii) the Q value, (iv) the energy dissipated in the first cycle, if initially the capacitor is charged to a p.d. of 10 V, and no current flows.

9. An inductance 10^{-4} H, a capacitance 10^{-10} F, a resistance $200\ \Omega$ and a sinusoidal oscillator of e.m.f. 1 V (peak value) are all connected in series. Deduce (i) the frequency of the natural damped oscillations of the circuit, (ii) the frequencies of charge resonance and current resonance and (iii) the Q value.

   Draw a rough sketch of the current in the circuit against oscillator frequency, indicating the maximum value of current, and the range of frequencies of particular interest. Draw a rough sketch of the phase difference between the e.m.f. of the oscillator, and (i) the current (ii) the p.d. across the capacitor, against frequency in both cases.

4 Mathematical description of wave motion

4.1 Waves in one dimension: the function $y = f(x - ct)$

Suppose we have a very long horizontal elastic string, originally at rest. Let us choose coordinate axes such that the x-axis is along the string, the y-axis vertically upwards, and the origin at some convenient point. The situation is shown in Fig. 4.1.

Suppose now that the string is set in motion by its being given a sudden flick at a point to the left of the origin. Experience tells us that this would result in a disturbance travelling down the string, the particles making up the string being momentarily displaced from their original positions. Let us assume that this disturbance takes place parallel with the y-axis (i.e. that is a true transverse wave) so that we may take the y-value of any point on the string as a measure of the disturbance of that point at a given instant of time. If a high-speed photograph is taken during the passage of the disturbance, the string will be seen to be distorted into a curve. This curve is referred to as the *wave profile*. We shall assume that this profile moves down the string with constant velocity c, and *without change of shape*, i.e. with no *dispersion*. Thus if we take two photographs at times t_1 and t_2, both would show the same profile, but in the second photograph the profile would be displaced along the

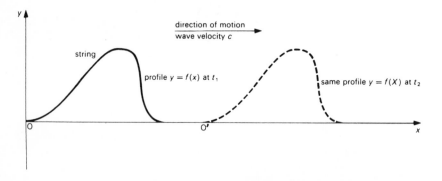

Figure 4.1 Waves on a long string.

string in the direction of propagation. Let us describe the shape of the profile at time t_1 by the function

$$y = f(x).$$

On the second photograph, taken at time t_2, let us mark the point O′ on the x-axis which is at the same position relative to the profile as the origin O was in the photograph taken at time t_1. Let distance along the x-axis referred to O′ be measured by the quantity X. Thus the shape of the profile referred to O′ at time t_2 is

$$y = f(X).$$

It is obviously not convenient to have our origin of coordinates moving along with the profile; we want to refer the profile to the fixed point O. Now since the profile is moving with constant c, the distance OO′ is $c(t_2 - t_1)$. Thus

$$X = x - c(t_2 - t_1).$$

The profile at time t_2 is then described by

$$y = f[x - c(t_2 - t_1)].$$

Finally, if our clock was started as the profile passed the point O, that is, $t_1 = 0$, then the profile at any subsequent time t is obtained by replacing the quantity $t_2 - t_1$ in this equation with the single quantity t, to give

$$\boxed{y = f(x - ct).} \tag{4.1}$$

This is an extremely important equation in the theory of wave motion. It completely defines a one-dimensional transverse wave of constant profile moving with constant velocity c along the positive direction of the x-axis. It is easy to show on the same lines, and it is left to the reader to do so, that a wave which is the same in all respects but moving in the opposite direction (i.e. along the direction of x decreasing) is given by

$$y = f(x + ct). \tag{4.2}$$

Equations (4.1) and (4.2) are examples of functions of two variables; to find y, we must know the values of the variables x and t (we must also know the form of the function f and the velocity c, but these do not change). Physically, this means that in order to find the disturbance suffered by a point on the string, we must specify not only its position along the string (x), but also the time (t) at which we wish to know the disturbance.

Neither (4.1) nor (4.2) is a completely general description of wave motion in one dimension, for each depends on a specific direction of propagation. We now look for a single expression which has the same generality for one-dimensional waves as

$$\frac{\mathrm{d}^2 x}{\mathrm{d}t^2} + \omega^2 x = 0 \tag{[2.1]}$$

has for one-dimensional vibrations.

We start by differentiating (4.1) (or (4.2) for the result is the same for both) to eliminate from it all reference to the function f and the direction of propagation, as follows. In equation (4.1), let us write

$$z = x - ct. \tag{4.3}$$

Then, differentiating $y = f(x - ct) = f(z)$ with respect to t, we obtain

$$\frac{\partial y}{\partial t} = \frac{df}{dz} \frac{\partial z}{\partial t}.$$

(The symbol ∂ indicates partial differentiation; for a full treatment of this subject see *Mathematical Methods for Mathematicians, Physical Scientists and Engineers* by J. Dunning-Davies, Ellis Horwood, 1981. Note, however, the d in df/dz since the function is one of z only.) But

$$\frac{\partial z}{\partial t} = -c \quad \text{(from 4.3)}$$

and so

$$\frac{\partial y}{\partial t} = -c \frac{df}{dz}. \tag{4.4}$$

Similarly,

$$\frac{\partial y}{\partial x} = \frac{df}{dz} \frac{\partial z}{\partial x},$$

but, as $\partial z / \partial x = 1$ (from 4.3), we have

$$\frac{\partial y}{\partial x} = \frac{df}{dz}. \tag{4.5}$$

Eliminating df/dz between (4.4) and (4.5) we obtain

$$\frac{\partial y}{\partial t} = -c \frac{\partial y}{\partial x}. \tag{4.6}$$

Now we repeat the same process, but starting with $y = f(x + ct)$. Let $x + ct = w$; this leads to

$$\frac{\partial y}{\partial t} = +c \frac{df}{dw} \tag{4.7}$$

and

$$\frac{\partial y}{\partial x} = \frac{df}{dw}. \tag{4.8}$$

Eliminating df/dw between (4.7) and (4.8) gives

$$\frac{\partial y}{\partial t} = +c\,\frac{\partial y}{\partial x}. \tag{4.9}$$

We see that equations (4.6) and (4.9), though very similar, are not identical; different results have been obtained for different directions of propagation. Let us see if we can eliminate all reference to direction of propagation by further differentiation. Differentiating (4.4) a second time with respect to t we have

$$\frac{\partial^2 y}{\partial t^2} = -c\,\frac{\partial}{\partial t}\left(\frac{df}{dz}\right)$$

$$= -c\,\frac{d}{dz}\left(\frac{df}{dz}\right)\frac{\partial z}{\partial t}.$$

But

$$\frac{d}{dz}\left(\frac{df}{dz}\right) = \frac{d^2 f}{dz^2}$$

and $\partial z/\partial t = -c$ (from 4.3). Therefore

$$\frac{\partial^2 y}{\partial t^2} = c^2\,\frac{d^2 f}{dz^2}. \tag{4.10}$$

Similarly, differentiating (4.5) with respect to x leads to

$$\frac{\partial^2 y}{\partial x^2} = \frac{d^2 f}{dz^2}. \tag{4.11}$$

Finally, eliminating $d^2 f/dz^2$ between (4.10) and (4.11) gives

$$\boxed{\frac{\partial^2 y}{\partial x^2} = \frac{1}{c^2}\,\frac{\partial^2 y}{\partial t^2}.} \tag{4.12}$$

If we repeat the same process, starting with

$$y = f(x + ct),$$

we obtain precisely the same final result as (4.12). This means that we have now obtained an equation which is completely independent of the direction of propagation. Equation (4.12) is an example of a *second-order partial differential equation*. It is known as the non-dispersive wave equation. The importance of this equation lies in its complete generality with regard to the form and direction of travel of waves which can be propagated in accordance with it. Examples of the occurrence of this equation in physics will be treated in Chapter 6 and its solutions will be discussed in detail in Chapter 5.

4.2 Harmonic waves

So far we have left the form of the function f in the equation

$$y = f(x - ct) \qquad [4.1]$$

completely arbitrary. In other words, our wave profile may have the shape of any continuous curve. It turns out that the simplest wave to treat analytically is one whose profile is a pure sine curve. We can express such a wave as

$$y = f(x - ct) = a \sin k(x - ct), \qquad (4.13)$$

where a and k are constants whose significances will appear shortly. Such a wave is known as a *sine wave*.

Suppose a wave described by equation (4.13) is propagated along a stretched elastic string of the kind described earlier in this chapter. How would a point on the string be disturbed due to the passage of the wave? We can answer this question by inserting the position of the point in question into (4.13). Let its position be x_1. Hence

$$y = a \sin k(x_1 - ct) \qquad (4.14)$$

or

$$y = -a \sin k(ct - x_1). \qquad (4.15)$$

This equation tells us how the disturbance or transverse position y of the point varies with time t. Note that y is now a function of the single variable t, since x has been given the constant value x_1. Equation (4.14) or (4.15) is therefore the equation of motion of the point at $x = x_1$.

We saw, in Chapter 2, that a point executing simple harmonic motion has the equation of motion

$$y = a \sin(2\pi ft + \varepsilon). \qquad [2.4]$$

Comparing (2.4) with (4.15) we see that these equations are really the same, except that the constants are differently arranged, so the point on the string will oscillate with simple harmonic motion. Furthermore, since we get the same form of equation (4.14) no matter what value of x is inserted, it follows than any point on the string along which a sine wave is propagated is caused to oscillate with simple harmonic motion. (For this reason, sine waves are also referred to as *harmonic* waves.) Thus all points along the string execute s.h.m. of the same amplitude and frequency but differing in phase. Points just ahead of x_1 will lag in phase while points just before x_1 will lead in phase.

A wave whose profile is that of a cosine function is very similar; sine and cosine functions have exactly the same form, the only difference between them being the point at which the origin is chosen. Since the choice of origin is always completely arbitrary, the first minus sign in (4.15) can be removed by a new choice of origin.

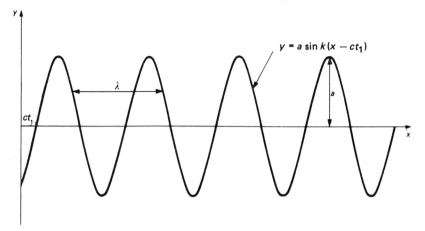

Figure 4.2 Graph of (4.13) for fixed *t*.

In order to get a complete picture of the motion of the point at $x = x_1$, we must compare corresponding terms in (4.15) and (2.4). We can now identify the quantity *a* in (4.15) with the amplitude of the motion caused by the wave. The wave itself is said to have amplitude *a*.

Comparing (2.4) and (4.15) we see that $2\pi f = kc$ or

$$k = \frac{2\pi f}{c}. \tag{4.16}$$

We thus have a physical meaning for *k* in terms of the frequency *f* of the oscillations caused by the wave, and of the wave velocity *c*. Further, since the period *T* of a simple harmonic motion is $1/f$, we can identify the period of the wave as $2\pi/kc$.

If, for equation (4.13), *y* is plotted against *x* for a given value of *t* (t_1 say), the curve shown in Fig. 4.2, which represents (essentially) a high-speed photograph of the string during the passage of a harmonic wave, is obtained. Since the sine function is periodic, the wave profile repeats itself after fixed intervals of *x*. The repeat distance is known as the *wavelength* and is designated by λ.

If we increase *x* by λ in equation (4.13) the value of *y* will, by definition, be unaltered, i.e.

$$y = a \sin k(x - ct) = a \sin k(\overline{x + \lambda} - ct).$$

It follows that $k\lambda = 2\pi$ or

$$\boxed{k = \frac{2\pi}{\lambda}.} \tag{4.17}$$

From (4.16) and (4.17) we see that

$$k = \frac{2\pi}{\lambda} = \frac{2\pi f}{c}.$$

which gives the extremely important result

$$c = f\lambda. \tag{4.18}$$

Thus the product of the frequency and the wavelength is equal to the speed of the wave. We are now in a position to rewrite (4.13) in a number of equivalent forms:

$$y = a \sin \frac{2\pi}{\lambda} (x - ct),$$

$$y = a \sin 2\pi \left(\frac{x}{\lambda} - ft \right),$$

$$y = a \sin 2\pi \left(\frac{x}{\lambda} - \frac{t}{T} \right).$$

If we now define the *wavenumber* σ as the number of wavelengths per metre, then $\sigma = 1/\lambda$, and

$$y = a \sin 2\pi(\sigma x - ft).$$

By a different initial choice of origin, we could equally well have arrived at the expression

$$y = a \sin 2\pi(ft - \sigma x). \tag{4.19}$$

Both representations of sine waves are commonly used, and they differ only in the choice of origin along the x-axis.

A mathematically more compact way of writing (4.19) is

$$y = a \sin(\omega t - kx) \tag{4.20}$$

where the *circular frequency* $\omega = 2\pi f$ as in section 2.2.1, and the *circular wavenumber* $k = 2\pi/\lambda$ as shown in (4.17) above.

4.3 Exponential representation of a harmonic wave

It can be shown from the elementary theory of complex numbers that

$$\exp i\theta = \cos \theta + i \sin \theta, \tag{4.21}$$

from which it follows that

$$\exp(-i\theta) = \cos\theta - i\sin\theta, \tag{4.21a}$$

where i is the imaginary quantity $\sqrt{(-1)}$. Here exp iθ is a complex quantity, which is expressed as the sum of a real part, cos θ, and an imaginary part, sin θ. In this notation cos θ may be referred to as the *real part* of exp iθ – abbreviated to Re(exp iθ) – and sin θ as the *imaginary part* of exp iθ – abbreviated to Im(exp iθ).

In this notation, equation (4.20) may be written as

$$y = a\sin(\omega t - kx) = \text{Im}[a\exp i(\omega t - kx)].$$

Finally, we drop the Im from the written expression since it will be understood, when a sine wave is expressed in this form, that it is the imaginary part of the expression that has physical meaning. Thus,

$$\boxed{y = a\exp i(\omega t - kx).} \tag{4.22}$$

Similarly, when we wish to treat a cosine wave in this way, the real part of (4.22) is implied.

The advantage of this procedure is that exponentials are much easier to handle mathematically than sines and cosines; they are easier to integrate, differentiate and sum as series. The procedure is as follows. We express our sine (or cosine) waves in exponential form; then we carry out our manipulation and take the imaginary (or real) part of the result as the quantity which is physically meaningful.

That the exponential and trigonometrical treatments yield identical results is demonstrated in the following simple example of wave addition. Suppose we have harmonic waves with identical amplitudes, frequencies and wavenumbers moving in opposite directions;

$$y_1 = a\sin(\omega t - kx)$$

and

$$y_2 = a\sin(\omega t + kx).$$

The sum of these is clearly

$$y = y_1 + y_2 = 2a\sin\omega t\cos kx. \tag{4.23}$$

If we now express the two waves as exponentials (the imaginary part being tacitly understood), the sum is

$$y = a\exp i(\omega t - kx) + a\exp i(\omega t + kx),$$

which, on factorising, becomes

$$y = a\exp i\omega t[\exp ikx + \exp(-ikx)]. \tag{4.24}$$

By eliminating sin θ between (4.21) and (4.21a) we see that

$$\cos \theta = \tfrac{1}{2}[\exp i\theta + \exp(-i\theta)].$$

Hence (4.24) becomes

$$y = 2a \exp i\omega t \cos kx.$$

Finally, writing

$$\exp i\omega t = \cos \omega t + i \sin \omega t$$

(by 4.21) and taking the imaginary part, we have

$$y = 2a \sin \omega t \cos kx,$$

which is the same as (4.23). The reader will hardly gain the impression from this example that the exponential representation saves labour, but it does demonstrate that the correct result is obtained. Several further examples in this book will, however, bring home the usefulness of this approach. Finally, in adding two waves together we have anticipated some of the content of Chapter 5, where the physical significance of wave addition is discussed fully.

4.4 Waves in two and three dimensions: wavefronts

4.4.1 Two-dimensional waves: straight-line wavefronts

So far, we have confined our attention to waves in one-dimensional media, of which the stretched elastic string is an example. We must now extend our theory first to two, and finally three, dimensions. A convenient example of waves in a two-dimensional medium is that of water waves in, say, a ripple tank. Let us imagine that the surface of the water in the ripple tank has been disturbed, for example by dropping a long stick so that it enters the water horizontally. A disturbance will proceed along the surface in the form of a straight-line crest which, for the purpose of the present argument, we shall assume moves with constant velocity, and without change of shape, in directions perpendicular to the stick.

We shall need two Cartesian coordinates x and y to specify the position of a point on the surface of the water, and we shall designate the disturbance, which in this case is the vertical displacement of a point on the surface from its undisturbed position, by ϕ.

Figure 4.3 shows schematically what we would see on a high-speed photograph of such a system taken from above, whilst Fig. 4.4 shows a vertical section in the direction in which the disturbance is travelling, which corresponds exactly to the one-dimensional wave we examined at the beginning of this chapter.

We meet now for the first time the concept of the wavefront; in the present case this is any continuous line joining points undergoing identical disturbance. Thus a crest is a wavefront, but so also is any other line joining points of equal disturbance. The importance of this concept resides in the fact that if we fix the state of disturbance

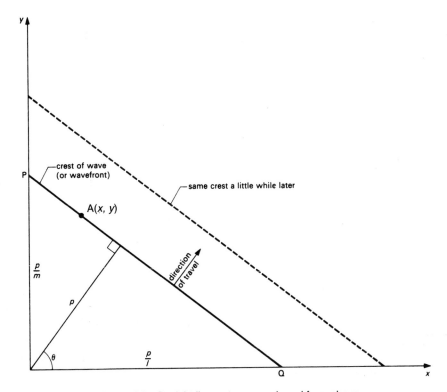

Figure 4.3 Straight-line water wave viewed from above.

as defining the wavefront, then the wavefront moves forward with the wave. In the present example the wavefront is a straight line, but we shall see later this is not always the case.

We must now look for a way of describing waves propagated with constant velocity and shape in two dimensions, corresponding to $f(x - ct)$ for one-dimensionl waves. What we do is to replace x in the one-dimensional equation by a quantity, containing both x and y, which is constant along a given wavefront. Suppose that a given wavefront has the instantaneous position given by the line PQ in Fig. 4.3. Now any point along this line will satisfy the equation

$$lx + my = p, \tag{4.25}$$

where, l, m and p are constants. (This is not the usual form of equation for a straight line used in coordinate geometry, but if we divide through by m and rearrange we get $y = -(l/m)x + (p/m)$, which is now in the more usual form $y = Ax + B$. The reason why we have used this apparently more complicated form will be seen when we discuss waves in three dimensions.)

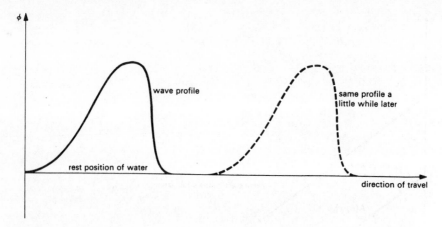

Figure 4.4   Vertical section through Fig. 4.3 in the direction of travel of the wave.

The meanings of the constants l, m and p will be seen from Fig. 4.3. Let p be the length of the perpendicular (or normal) from the origin to the line PQ. If this normal, which is in the direction of travel of the wave, makes an angle θ with the Ox-axis, then $\cos \theta = l$ and $\sin \theta = m$. Thus l is the cosine of the angle the normal makes with the Ox-axis and m is the cosine of the angle the normal makes with the Oy-axis; l and m are known as the *direction cosines* of the normal. Clearly

$$l^2 + m^2 = 1.$$

Thus (4.25) expresses the equation of a straight line in terms of the direction cosines of the line normal and the perpendicular distance from the origin to the line.

Since the wave is travelling in the direction of p, and since the value of p is constant for a wavefront at a given time, the function $\phi = f(p - ct)$ is that which describes the wave. As

$$p = lx + my, \qquad\qquad\qquad [4.25]$$

we can write the wave function finally as

$$\phi = f(lx + my - ct). \qquad\qquad\qquad (4.26)$$

Thus (4.26) specifies a wave in two dimensions, of constant profile and constant velocity, with wavefronts which are straight lines, moving in the direction having direction cosines (l, m). Similarly it can be shown that

$$\phi = f(lx + my + ct)$$

represents a wave of identical profile moving in the opposite direction. If we eliminate the functional form f, and reference to the direction of propagation, from (4.26) by differentiation, along similar lines to the one-dimensional case, we obtain the partial differential equation

$$\frac{\partial^2 \phi}{\partial x^2} + \frac{\partial^2 \phi}{\partial y^2} = \frac{1}{c^2} \frac{\partial^2 \phi}{\partial t^2}.$$

(4.27)

This is the partial differential equation governing straight-line wave propagation in two dimensions; but in fact it covers wavefronts of any shape.

4.4.2 Two-dimensional harmonic waves: vector representation

A harmonic wave in two dimensions is typified by a water wave whose vertical section in the direction of travel is a sine curve. We may therefore mathematically describe such a wave, whose wavefronts are straight lines, by the equation

$$\phi = a \sin \omega (lx + my - ct)/c.$$

(4.28)

The quantity ω/c is obtained by the same process as that which led to equation (4.16) earlier in this chapter, whilst a represents the amplitude of the wave, as before. A high-speed photograph of water waves of this kind taken from above would be characterised by a family of equispaced parallel crests, the perpendicular distance between adjacent crests being the wavelength λ.

Two adjacent crests, PQ and P'Q', are shown in Fig. 4.5. (Actually PQ and P'Q' need not be crests, but any pair of adjacent identical wavefronts.) Let us define unit vectors **i** and **j** parallel to Ox and Oy respectively, and let the unit vector along the direction of wave travel (i.e. the perpendicular to the wavefront) be **u** as shown. Take any point $A(x, y)$ with vector position **r** on the wavefront PQ; we can express the vector **r** in terms of its components as follows:

$$\mathbf{r} = x\mathbf{i} + y\mathbf{j}.$$

Similarly we can express the unit vector **u** as

$$\mathbf{u} = l\mathbf{i} + m\mathbf{j},$$

where l and m are the direction cosines of **u**. The scalar product of **r** and **u** is

$$\mathbf{r}.\mathbf{u} = (x\mathbf{i} + y\mathbf{j}).(l\mathbf{i} + m\mathbf{j})$$

$$= lx + my = p.$$

(4.29)

The equation $\mathbf{r}.\mathbf{u} = p$ is the vector equation for a straight line, and it enables us to write (4.28) in the vector form,

$$\phi = a \sin \frac{\omega}{c} (\mathbf{r}.\mathbf{u} - ct).$$

(4.30)

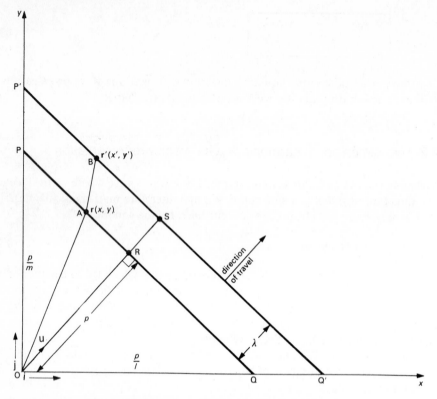

Figure 4.5 Vector representation of straight-line wave.

Now PQ and P'Q' are identical wavefronts, so that if we replace **r** in (4.30) by the vector position of *any* point on P'Q' it follows from the definition of a wavefront that the value of ϕ must remain unaltered. Let B [vector position **r**', coordinates (x', y')] be such a point. Thus

$$\phi = a \sin \frac{\omega}{c}(\mathbf{r.u} - ct) = a \sin \frac{\omega}{c}(\mathbf{r'.u} - ct). \tag{4.31}$$

But **r.u** is the distance OR (Fig. 4.5) and **r'.u** is the distance OS. Thus

$$\mathbf{r'.u} - \mathbf{r.u} = RS = \lambda. \tag{4.32}$$

since λ is the perpendicular distance between PQ and P'Q'.

Substituting from (4.32) into (4.31) we have

$$\phi = a \sin \frac{\omega}{c}(\mathbf{r'.u} - ct) = a \sin \frac{\omega}{c}[(\mathbf{r.u} + \lambda) - ct].$$

Clearly

$$\frac{\omega}{c} \lambda = 2\pi$$

and

$$\frac{\omega}{c} = \frac{2\pi}{\lambda},$$

so we can rewrite (4.30) in the form

$$\phi = a \, \sin\!\left(\frac{2\pi \mathbf{r}.\mathbf{u}}{\lambda} - \omega t\right).$$

We now extend the earlier definition of the wavenumber $\sigma \, (= 1/\lambda)$ so that it becomes the *wavevector* $\mathbf{\sigma}$. The direction of $\mathbf{\sigma}$ is perpendicular to the wavefront (that is, parallel to the unit vector $\mathbf{u}$) whilst the magnitude $|\mathbf{\sigma}|$ is $1/\lambda$, as before. Hence, $\mathbf{u}/\lambda = \mathbf{\sigma}$ and the above equation becomes

$$\phi = a \, \sin(2\pi\mathbf{\sigma}.\mathbf{r} - \omega t) \tag{4.33}$$

or, writing $\omega = 2\pi f$,

$$\phi = a \, \sin 2\pi(\mathbf{\sigma}.\mathbf{r} - ft) \tag{4.34}$$

A mathematically more compact form is obtained by introducing the *circular wavevector* $\mathbf{k} = 2\pi\mathbf{\sigma}$, so that (4.33) becomes

$$\phi = a \, \sin(\mathbf{k}.\mathbf{r} - \omega t) \tag{4.35}$$

This is the vector representation of a harmonic wave we shall most commonly use, particularly in its exponential form

$$\boxed{\phi = a \, \exp i(\mathbf{k}.\mathbf{r} - \omega t).} \tag{4.36}$$

4.4.3 Plane waves in three dimensions

Equations (4.34) to (4.36) can be taken over, without change of form, into three dimensions. The three-dimensional equivalent of the straight line in two dimensions is the plane, so that the straight-line wavefronts of the previous section become plane wavefronts in three dimensions, and now $\mathbf{r}$ specifies a point in three-dimensional space.

If we wish to specify a three-dimensional plane wave in Cartesian form, we can extend the two-dimensional equation (4.26) to

$$\phi = f(lx + my + nz - ct), \tag{4.37}$$

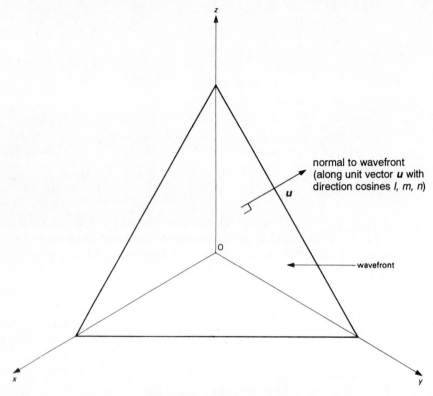

Figure 4.6 A plane wavefront.

where (l, m, n) are the direction cosines of the normal to the plane wavefront (Fig. 4.6). The partial differential equation, which is the three-dimensional equivalent of (4.27), is

$$\frac{\partial^2 \phi}{\partial x^2} + \frac{\partial^2 \phi}{\partial y^2} + \frac{\partial^2 \phi}{\partial z^2} = \frac{1}{c^2} \frac{\partial^2 \phi}{\partial t^2}, \tag{4.38}$$

which is easily verified by differentiating (4.37).

It is convenient to write (4.38) in the form

$$\nabla^2 \phi = \frac{1}{c^2} \frac{\partial^2 \phi}{\partial t^2},$$

where ∇^2 (referred to in speech as 'del squared') is an abbreviation of

$$\frac{\partial^2}{\partial x^2} + \frac{\partial^2}{\partial y^2} + \frac{\partial^2}{\partial z^2},$$

and is often referred to as the Laplacian.

4.5 Circular and spherical wavefronts

Straight-line wavefronts (in two dimensions) and plane ones (in three) are by no means the only wavefronts encountered in physics, though they are the simplest to treat mathematically. If we have a point source of disturbance in two dimensions, such as a pebble dropping on to a previously still surface of water, the resulting wavefronts are of circular form since the disturbance proceeds outwards from the point of entry with equal speed in all directions. In three dimensions, the wavefronts arising from a point source of disturbance are spherical in form.

Problems

1. Which of the following disturbances represent a travelling wave in one dimension? In each case, what is the speed of the wave, and in which direction is it travelling? (Ignore the fact that some of these are unrealistic, since the disturbance may tend to infinity.)

(i) $y = (3x - 4t)^2$
(ii) $y = x^2 t^2$
(iii) $y = e^{-\alpha x} e^{i\omega t}$
(iv) $y = e^{ikx} e^{i\omega t}$
(v) $y = \exp[-\alpha(2x - t)^2]$
(vi) $y = \sin 3t \sin 4x$
(vii) $y = \sin(x + 2t) + \sin(2x + 3t)$
(viii) $y = \sin(x + 2t) + \sin(2x + 4t)$

2. Use a calculator to plot one cycle of the disturbance $y = \sin 2\pi(0.2x - 5t)$ at (i) $t = 0$, (ii) $t = 1/10$ of the period and (iii) $t = 1/4$ of the period. The progressive nature of the wave should be obvious from the graphs. What are the wave speed, wavelength, frequency and circular wavenumber?

3. A travelling sinusoidal wave of frequency f moves in the positive direction of x with speed c. P and Q are points on the x-axis having coordinates x and $x + dx$ respectively. What is the phase difference between the disturbances at P and Q? Does that at Q 'lead' or 'lag behind' that at P?

4. The human ear can perceive sounds over a frequency range 30 Hz to 15 kHz. The speed of sound in air is 340 m s^{-1}. What are the wavelengths at these extremities?

5. The speed of electromagnetic waves in free space is 3×10^8 m s^{-1}. Calculate (i) the wavelength of a very high frequency (VHF) radio signal at 100 MHz, (ii) the wavelength of waves emitted by a wire connected to the domestic mains supply (50 Hz), (iii) the frequency of visible light of wavelength 500 nm and (iv) the frequency of X-rays of wavelength 0.1 nm.

6. Analyse the problem of 'beats' (section 2.3) using complex wave notation, letting the disturbances be $a \exp i\omega_1 t$ and $a \exp i\omega_2 t$. Show that the resulting wave has circular frequency $\frac{1}{2}(\omega_1 + \omega_2)$, and that the square of the amplitude oscillates with circular frequency $|\omega_1 - \omega_2|$.

7.

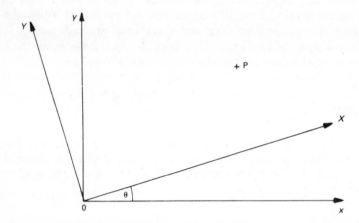

The coordinates of a point P are (x, y) when measured in relation to the axes (x, y) and (X, Y) in relation to the axes (X, Y) rotated through θ as shown. Show that $X = x \cos \theta + y \sin \theta$, and $Y = -x \sin \theta + y \cos \theta$. Hence show that equation (4.28) describes a plane wave moving, as expected, along the X-axis.

8. Show that a possible solution of the two-dimensional wave equation is $\phi = A \sin k_1 x \sin k_2 y \sin \omega t$. What are the significances of k_1 and k_2? What is the wave speed?

5 Superposition of waves: initial and boundary conditions

5.1 Introduction

This chapter is concerned with the general solution of

$$\frac{\partial^2 y}{\partial t^2} = c^2 \frac{\partial^2 y}{\partial x^2} \tag{5.1}$$

and the physical implications of the solution. We saw in Chapter 4 how (5.1) is obtained from the wave $y = f(x - ct)$. This means that $y = f(x - ct)$ is a solution of (5.1); we showed that $y = g(x + ct)$ is also a solution of the equation. (We call the function here g since it need not be the same function as the previous one.) Further, since the original wave equation is a *linear* partial differential equation, *any* linear sum of these two solutions is itself a solution. That is to say, a perfectly good solution of the wave equation (5.1) is

$$\boxed{y = f(x - ct) + g(x + ct).} \tag{5.2}$$

In fact, it can be shown (but we shall not do so) that (5.2) is the *general solution* to the wave equation. This is commonly referred to as D'Alembert's solution.

There are two very striking things about this solution. The first is that it gives us a good deal of information about possible waves in a one-dimensional system governed by the wave equation (5.1). It tells us that either or both directions of wave propagation are possible (a superimposition of waves in both directions being the general case, and special cases of one direction only being when either f or g is zero). Static equilibrium is also consistent with the wave equation, since f and g can be zero. There is, of course, no wave at all in this trivial case, but it would be strange if the wave equation explicitly ruled out the possibility of the system existing in a state of static equilibrium. Another important fact that (5.2) tells us is that the speed has to be the same for both directions of propagation. As we will see in the next chapter, the value of c is determined solely by the physical properties of the wave system under consideration.

The second important feature about the solution (5.2) concerns the *lack* of information it gives. Not only can the waves travel (with equal speed) in either direction, but they can be of any form whatsoever, since there is no effective restriction on the nature of the functions f and g. So it becomes apparent that a waveform of *any* shape may be propagated in accordance with the wave equation. This, after all, ties up with our everyday experience. The air around us will transmit sound waves in accordance with the three-dimensional equivalent of the wave equation (5.1), and shows absolutely no preference to any particular signal. It will transmit Beethoven, Bartok and the Beatles with equal ease (and, indeed, at the same speed!); this enviable versatility is the prominent feature of D'Alembert's solution.

5.2 Superposition of waves

Suppose that a room is filled with music. The air is vibrating with a comparatively large amplitude and a fairly complicated waveform. It would therefore seem to be remarkable that one can simultaneously hear a much quieter sound, for instance the rattling of a teacup, with perfect clarity and no distortion. It sounds exactly the same as if the music were absent. The two sounds are evidently propagated independently, and do not interact with each other. The same is true for light propagation – two torch beams can pass through each other with no interaction.

The reason for this is quite straightforward, and to be found via the wave equation

$$\frac{\partial^2 y}{\partial x^2} = \frac{1}{c^2} \frac{\partial^2 y}{\partial t^2}. \qquad [4.12]$$

We have taken the one-dimensional equation for simplicity, but the extension of the argument to three dimensions is straightforward.

Suppose that the music alone were being played. Then its waveform y_1 would follow

$$\frac{\partial^2 y_1}{\partial x^2} = \frac{1}{c^2} \frac{\partial^2 y_1}{\partial t^2}. \qquad (5.3)$$

Similarly, for the rattling teacup with no music,

$$\frac{\partial^2 y_2}{\partial x^2} = \frac{1}{c^2} \frac{\partial^2 y_2}{\partial t^2}. \qquad (5.4)$$

Adding these two equations gives

$$\frac{\partial^2}{\partial x^2} (y_1 + y_2) = \frac{1}{c^2} \frac{\partial^2}{\partial t^2} (y_1 + y_2) \qquad (5.5)$$

which tells us that $y_1 + y_2$ is a *possible solution of the wave equation*, with the total displacement equal to the sum of the two displacements due to each of the sources alone.

This is the *principle of superposition* of waves. The reason why it comes about is that the wave equation is a *linear* differential equation. This means that if y in (4.12) were doubled, each term would also be doubled. (An example of a non-linear wave equation will be considered in section 6.11.) The principle is clearly true for the superposition of any number of waves – the net disturbance is the sum of the disturbances due to each of the sources acting independently of the others.

An example of superposition has already been encountered in d'Alembert's solution (5.2), and we shall meet numerous other examples in the course of this book.

5.3 Standing (stationary) waves

A particularly important instance of superposition is that of two sinusoidal waves of the same frequency, travelling in opposite directions. This may arise for instance if a sinusoidal sound wave is directed towards, and reflected from, a wall. The incident and reflected waves are superimposed. Another example concerns a vibrating string fixed at both ends, to be considered in detail presently.

Suppose in the first instance that the amplitudes are equal. The individual waves are expressible as $y_1 = a \sin(\omega t - kx)$ for the left-to-right wave and $y_2 = a \sin(\omega t + kx)$ for the right-to-left wave. (By a judicious choice of the origins of t and x, no extra phase factors are necessary.) The total disturbance is, since $\sin A + \sin B = 2 \sin \frac{1}{2}(A + B) \cos \frac{1}{2}(A - B)$,

$$y = y_1 + y_2 = 2a \sin \omega t \cos kx. \tag{5.6}$$

Note that this function is not of the form $f(x \pm ct)$, but the product of two terms, one involving only t as a variable and the other involving only x. Consider now what happens at the point $x = 0$. Since here $\cos kx = 1$, (5.6) becomes

$$[y]_{x=0} = 2a \sin \omega t,$$

so the total displacement is sinusoidal with amplitude $2a$. (The left-hand side is a notation meaning 'the value of y when $x = 0$'.)

Consider now the point one-eighth of a wavelength away from the origin, so $x = \lambda/8$. Since $k = 2\pi/\lambda$, we see that $kx = \pi/4$, so that $\cos kx = 1/\sqrt{2}$. It follows that

$$[y]_{x=\lambda/8} = \sqrt{2}a \sin \omega t$$

At this point, the displacement varies with time in just the same manner as that at $x = 0$ because of the $\sin \omega t$ term, but the amplitude is reduced from $2a$ to $(\sqrt{2})a$.

Let us proceed further, and consider what happens at the point one-quarter of a wavelength from the origin, $x = \lambda/4$. kx is now $\pi/2$, therefore $\cos kx = 0$, so

$$[y]_{x=\lambda/4} = 0.$$

Therefore *the displacement at this point is zero at all times*. In fact, there is a series of such points, namely those for which $\cos kx = \cos(2\pi x/\lambda) = 0$, or

$$\frac{2\pi x}{\lambda} = \pm\frac{\pi}{2}, \pm\frac{3\pi}{2}, \pm\frac{5\pi}{2}, \ldots,$$

$$= (2n + 1)\frac{\pi}{2},$$

where $n = \ldots, -2, -1, 0, 1, 2, \ldots$, i.e.

$$x = (2n + 1)\frac{\lambda}{4}.$$

These points of no displacement are called *nodes*. Since any two adjacent nodes are a distance $\frac{1}{2}\lambda$ apart, half-way between the nth and $(n + 1)$th nodes the value of x will be given by

$$x = (2n + 1)\tfrac{1}{4}\lambda + \tfrac{1}{4}\lambda$$

$$= (2n + 2)\tfrac{1}{4}\lambda$$

$$= (n + 1)\tfrac{1}{2}\lambda.$$

For this value of x,

$$\cos kx = \cos\frac{2\pi x}{\lambda} = \cos[(n + 1)\pi] = \pm 1.$$

Such points will therefore oscillate in accordance with the equation

$$y = \pm 2a \sin \omega t.$$

No point can have a greater amplitude than this (since unity is the maximum value of the cosine function). Thus, half-way between each adjacent pair of nodes we have a point of maximum vibration. These points are known as *antinodes*.

Figure 5.1 shows the wave at different times. It is seen that there is apparently no wave travelling along the x-axis in either direction, the two constituent travelling waves in this case adding up to give a stationary effect. Such a superposition is called a *standing wave* or *stationary wave*, and is of great importance, as we shall see later.

We repeat an important mathematical point concerning standing waves. We saw that the standing waves in this particular case are described by (5.6), i.e.

$$y = 2a \sin \omega t \cos kx.$$

This is no longer of the form $y = f(x \pm ct)$ but is, rather, the *product* of two functions, $\cos(kx)$, which is a function of the position x only, and $2a \sin \omega t$, which is a function of the time t only. Standing waves are always described in this way, and, in general, we can write

$$y = F_1(x)F_2(t),$$

where $F_1(x)$ and $F_2(t)$ are, respectively, functions of position and time only. Whenever we see a wave represented by an equation of this type, we know at once that a

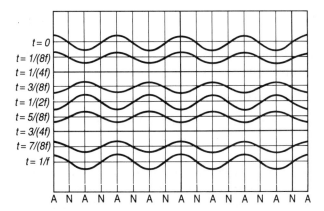

Figure 5.1 Standing-wave patterns at different times. N denotes a node, A an antinode.

standing wave is indicated. It is arguable whether this situation is, indeed, correctly describable as wave motion since there is no net flow of energy in any direction.

Finally, we consider briefly what happens if the two waves travelling in opposite directions have unequal amplitudes a_1 and a_2. The antinodes will have the maximum possible amplitude, which is now $a_1 + a_2$, and the nodes will have the minimum amplitude of $|a_1 - a_2|$, which is not zero in this case. Intermediate points will vibrate with intermediate amplitudes.

The reader should now view the demonstration on the available computer disk, which illustrates two sinusoidal waves of equal amplitude travelling in opposite directions, and the resulting standing wave. The amplitude of one of the waves can be reduced to zero. If this is done gradually, the wave looks more and more like a travelling wave.

5.4 Initial conditions

Although the solution (5.2) of the wave equation (5.1) is very general, this does not imply that the solution under any circumstances is so general. We shall now examine in some detail the factors which restrict the generality of the solution, and to illustrate this we shall take the case of a stretched string.

Consider a very long stretched string, so long that it can be thought of as being of infinite length. We shall see in the next chapter that any transverse wave propagated along the string must have a velocity $\sqrt{(T/\mu)}$ where T and μ are the tension, and mass per unit length of the string, respectively. Also, from the previous section, a wave of any shape may be propagated, subject to certain conditions, the most obvious one being that the shape must be represented by a *continuous* function – any discontinuity would correspond to a break in the string which would certainly not be propagated along its length.

Suppose initially we make the string take a certain shape, which we can describe by the function $\phi(x)$, and suppose that at time $t = 0$ we release the string. A wave will be propagated along the string. We already know that the general shape of the string at any subsequent time t is given by

$$y = f(x - ct) + g(x + ct). \tag{5.7}$$

We must interpret this equation in the light of our knowledge that at time $t = 0$

$$[y]_{t=0} = \phi(x). \tag{5.8}$$

Equation (5.8) states the initial position of every particle of the string; the left-hand side is merely a conventional shorthand for the value of the function y at time $t = 0$. Since we are constraining the string until time $t = 0$, at which time we are 'letting go', the initial velocities of all particles in the string are zero. That is to say,

$$\left[\frac{\partial y}{\partial t}\right]_{t=0} = 0, \text{ for all } x. \tag{5.9}$$

The two equations (5.8) and (5.9) describe what are known as *initial conditions* which, as we shall now see, restrict the generality of (5.7). By making t zero in (5.7) we get

$$[y]_{t=0} = f(x) + g(x).$$

But this, from (5.8), is equal to $\phi(x)$, i.e.

$$f(x) + g(x) = \phi(x). \tag{5.10}$$

Let us now return to (5.7) and partially differentiate it with respect to t:

$$\frac{\partial y}{\partial t} = \frac{\partial}{\partial t} f(x - ct) + \frac{\partial}{\partial t} g(x + ct)$$

$$= \frac{df(x - ct)}{d(x - ct)} \frac{\partial(x - ct)}{\partial t} + \frac{dg(x + ct)}{d(x + ct)} \frac{\partial(x + ct)}{\partial t}$$

$$= -c \frac{df(x - ct)}{d(x - ct)} + c \frac{dg(x + ct)}{d(x + ct)}.$$

At time $t = 0$ this reduces to

$$\left[\frac{\partial y}{\partial t}\right]_{t=0} = -c \frac{df(x)}{dx} + c \frac{dg(x)}{dx}.$$

By cross-multiplying, we see that this becomes

$$\frac{1}{c}\left[\frac{\partial y}{\partial t}\right]_{t=0} dx = -df(x) + dg(x),$$

and by integrating with respect to x we obtain

$$\frac{1}{c}\int\left[\frac{\partial y}{\partial t}\right]_{t=0}dx = -f(x) + g(x). \tag{5.11}$$

If we now subtract (5.11) from (5.10) we obtain

$$2f(x) = \phi(x) - \frac{1}{c}\int\left[\frac{\partial y}{\partial t}\right]_{t=0}dx \tag{5.12}$$

and if we add together (5.11) and (5.10) we obtain

$$2g(x) = \phi(x) + \frac{1}{c}\int\left[\frac{\partial y}{\partial t}\right]_{t=0}dx. \tag{5.13}$$

These last two equations define the general functions f and g of D'Alembert's solution in terms of the initial conditions we have imposed on one problem. In the particular case we are considering, we have assumed that the initial velocities of all particles of the string are zero, as stated in (5.9). We can immediately see that this effects a considerable simplification in (5.12) and (5.13) by getting rid of both integrals, and the equations become

$$f(x) = \tfrac{1}{2}\phi(x)$$

and

$$g(x) = \tfrac{1}{2}\phi(x).$$

Substituting these back into equation (5.7) we get

$$y = \tfrac{1}{2}\phi(x - ct) + \tfrac{1}{2}\phi(x + ct). \tag{5.14}$$

To illustrate the meaning of (5.14), let us consider a particular initial shape $\phi(x)$, illustrated in Fig. 5.2. Here the uniform string, which is under tension T, is bent into the shape shown, by three pegs A, B and C. We can imagine the horizontal axis as x and the vertical as $\phi(x)$. The pegs are all removed at the same instant (time $t = 0$). What is the subsequent behaviour of the string? First of all, (5.14) very properly tells us that, at time $t = 0$, y is indeed equal to $\phi(x)$, the two terms on the right-hand side each contributing half. But at a later time the two halves will not be, as they were

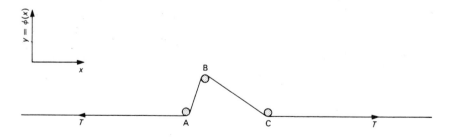

Figure 5.2 An example of initial conditions in a long string.

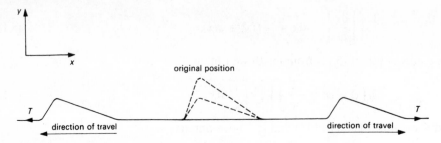

Figure 5.3 Appearance of the string of Fig. 5.2 at a later time.

at the beginning, coincident in space. In fact the first term in (5.14) represents the initial shape (scaled down by a factor of two) travelling to the right with speed c, while the second term represents the same shape (similarly scaled down) travelling to the left with the same speed. At a subsequent time, therefore, the string would look as in Fig. 5.3.

As time goes on, the two disturbances in the string proceed away from each other at a constant relative velocity $2c$.

5.5 Boundary conditions

5.5.1 Perfect reflection at a rigid boundary

Initial conditions specify the displacement, or rate of change of displacement, or some other time derivative of displacement of every particle of the system at a *particular instant* of time, and we have seen that they reduce the generality of the solution of the wave equation (5.1). In this section we are going to investigate the effect of what are known as *boundary conditions*, which occur in problems where we impose a restriction on what happens at a *particular* position in the system as time goes on. These two different types of conditions are rather similar; in the one case we specify what happens at all values of x for a particular value of t, and in the other case we specify what happens at all values of t for a particular value of x.

To illustrate the ideas involved in boundary conditions, let us refer once more to the case of the uniform, tensioned string. Suppose we have such a string stretching from $x = 0$ to $x = \infty$. At the point $x = 0$ the string is firmly clamped so that it cannot move. Since the transverse displacement of a point at distance x from the origin at time t is $y(x, t)$, as before, the restriction placed by the clamping on the possible wave motions of the string can be stated as

$$y(0, t) = 0 \tag{5.15}$$

for all time t. Equation (5.15) is a *boundary condition* that we are imposing in this particular problem.

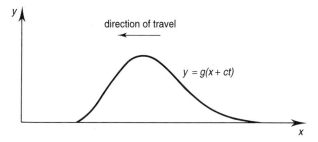

Figure 5.4 Transverse wave on a clamped string.

Now suppose we are able to propagate a transverse wave of the form shown in Fig. (5.4), and described by a function $g(x, t)$, along the string towards the origin. Since the disturbance is travelling in the negative direction of x, the wave must be described by $y = g(x + ct)$. What happens when the disturbance arrives at the clamp at $x = 0$? A simple experiment can help us here. A length of cord (a clothes line would do nicely) is firmly tied at one end to some fixed object and the other end is held fairly taut. If the held end is now given a sudden single flick, it is seen that the resulting disturbance travels down the cord to the tied end where it turns round and retraces its path back towards the hand-held end of the string.

What can we tell about this return wave? It must travel with the same speed as the outward-going wave since, as we have stated, the speed is determined entirely by the tension, and mass per unit length, of the string. We shall make no assumptions about the shape of the return wave, and shall describe it by the function $y = f(x - ct)$. As the incident wave reaches the point of clamping at $x = 0$ it can cause no transverse motion of the string, because the string is not free to move. It follows, therefore, that the return wave is one of such a shape as to cancel out the lateral displacement of the incident wave at $x = 0$. In other words, the displacement of the incident wave at $x = 0$ is exactly equal and opposite to the displacement there due to the return wave at all times, so that when we add the two displacements together at $x = 0$ the result is always zero. The reader should view the demonstration on the disk available.

The diagrams in Fig. 5.5 show the incident wave at various stages after its arrival at the clamped point $x = 0$. With the knowledge that the return wave has to be of such a shape so as to cancel the displacement at $x = 0$ due to the incident wave, we can build up the shape of the return wave as shown, bearing in mind that both waves have the same speed. Since the displacement due to this particular incident wave is upwards, the displacement due to the return wave must be downwards. If we follow the sequences in Fig. 5.5 through, we see that the shape of the wave has also been reversed in the x-direction. In other words, the original wave has been turned back-to-front and upside down.

To see this mathematically, we can represent the process as follows. When we insert the boundary condition (5.15) into (5.2) for $x = 0$, we obtain

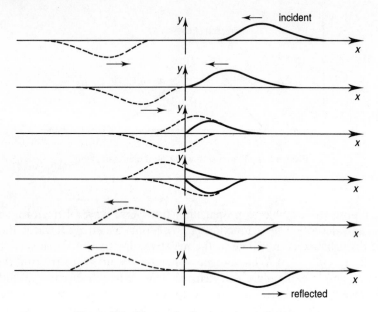

incident

reflected

Figure 5.5 Wave reflection in a clamped string.

$$[y]_{x=0} = f(-ct) + g(ct) = 0.$$

Thus

$$f(-ct) = -g(ct).$$

So, replacing the argument $-ct$ in the above equation by $x - ct$, we find that the relationship between f and g for this case is

$$f(x - ct) = -g(ct - x).$$

Thus, from equation (5.2)

$$y = g(x + ct) - g(ct - x). \tag{5.16}$$

The implication of equation (5.16) is that if we propagate a wave described by the function $g(x + ct)$ down the string, a second wave represented by $-g(ct - x)$ develops because of the restriction imposed at the origin. How is this latter function related to the original one? To understand this, we note that $g(x - ct)$ would be of exactly the same shape as $g(x + ct)$ but travelling in the opposite direction. It can also be noticed that $g(ct - x)$ is the same as $g[-(x - ct)]$, so we are faced with the question of the relationship of $g(X)$ to $g(-X)$. A little thought will enable one to see that these functions are mirror images of one another, the line of reflection being the y-axis (Fig. 5.6b). Finally we note that there is a minus sign in front of $g(ct - x)$. This, of course, implies an additional reflection of the function about the x-axis (Fig. 5.6c).

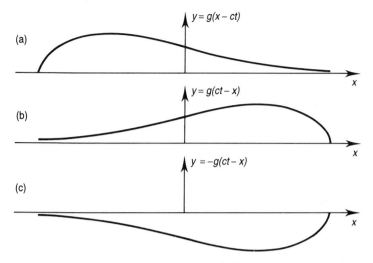

Figure 5.6 Illustration of the relationship between the functions $g(x - ct)$, $g(ct - x)$ and $-g(ct - x)$.

Let us now summarise the consequences of the imposition of this boundary condition. The original wave incident on the boundary (namely $g(x + ct)$) is reflected back with an equal speed from the origin. Moreover, the reflected wave is of the same shape as the incident wave, except that the shape is reversed left-to-right and has been turned upside down. This reflection process is very common in nature and, although we have taken as an example a particularly simple boundary condition, a great variety of such conditions has just this effect, that of producing a reflected wave. The details vary from one situation to another, but in most problems where the physical properties of the medium change, some form of reflection always takes place. It is the boundary condition at the surface of a mirror that causes it to reflect light waves, and it is a rather more complicated boundary condition at the bank of a river which causes the bank to reflect surface water waves. We shall examine several other kinds of boundary conditions later in the book.

5.5.2 Partial reflection

We have seen what happens when an incident wave arrives at a point on a stretched string which is securely clamped. In the ideal situation described, the profile is reflected without change of shape (apart from the inversion process). In this treatment, we implied that all energy in the incident wave appeared in the reflected wave – that is, there was no energy loss at the clamp. In practice, this would not, of course, be the case. Energy would be lost at the clamp (as well as during the progress of the wave along the string) and this would result in a change of amplitude in the reflected wave. The process becomes quite complicated to treat mathematically and this is why we have made the rather idealised assumptions in the previous section.

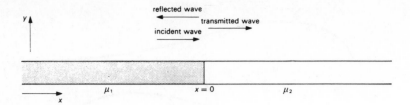

Figure 5.7 The junction of two stretched elastic strings. The arrows indicate the direction of travel of the transverse waves.

We will now go on to see what happens when the physical characteristics of the string change at some point. Suppose, for example, we have two stretched strings, both very long but of different mass per unit length, joined together as shown in Fig. 5.7. Let the join be at the point $x = 0$ and let the left- and right-hand strings have masses per unit length μ_1 and μ_2 respectively. Let us assume that a harmonic wave is sent along the left-hand string in the direction from left to right. We will use the exponential representation for harmonic waves which we discussed in section 4.3.

Let the incident wave be

$$y_i = A_i \exp i(\omega_i t - k_i x). \tag{5.17}$$

Here, the suffix i indicates the incident wave, A_i is the amplitude (a real number); ω_i and k_i are, respectively, the circular frequency and the circular wavenumber of the wave.

What happens when this wave arrives at the junction at $x = 0$? Let us assume that part of the disturbance is transmitted into the second string, but that the other part is reflected back along the left-hand string. This is not begging the question as might at first appear. If, in fact, there is no wave reflected or no wave transmitted, our analysis will show this. We may, therefore, represent the transmitted and reflected waves quite generally as follows:

$$y_t = A_t \exp i(\omega_t t - k_t x), \tag{5.18}$$

$$y_r = A_r \exp i(\omega_r t + k_r x), \tag{5.19}$$

where the suffixes t and r refer to the transmitted and reflected waves respectively (see Fig. 5.7). If if turns out that there are phase changes at the junction, A_t and A_r will prove to be complex quantities (or negative if the phase change is π).

Now by the principle of superposition, the total disturbance in the left-hand string will be the superposition of the incident and reflected waves. We can write this as $y_1 = y_i + y_r$. The total disturbance in the right-hand string is just the transmitted wave, so $y_2 = y_t$.

Let us now examine the physical conditions which govern the values of y_1 and y_2 at the junction. Firstly, if the string does not come apart, the values of y_1 and y_2 must always be the same, so

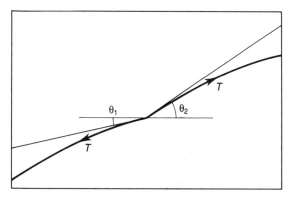

Figure 5.8 String with a discontinuity in $\partial y/\partial x$.

$$[y_1]_{x=0} = [y_2]_{x=0}$$

or

$$[y_i + y_r]_{x=0} = [y_t]_{x=0}. \tag{5.20}$$

The second condition is not so obvious. The transverse force acting at a point on a stretched string under tension T along which waves are travelling is $T \sin \theta \approx T(\partial y/\partial x)$, since θ is small (Fig. 5.8). Now this latter quantity must be continuous along the string. Suppose for a moment that this is not so. Then for a given point on the string there would be a transverse force $T \sin \theta_1 \approx T[\partial y/\partial x]_1$, say, due to the parts of the string to the left of the point, and a force $T \sin \theta_2 \approx T[\partial y/\partial x]_2$, say, due to the parts of the string to the right of the point. So the net force on the point on the string would be $T\{[\partial y/\partial x]_2 - [\partial y/\partial x]_1\}$. But this force must be zero, otherwise infinite acceleration would be suffered by the infinitesimal mass of the point. Therefore $[\partial y/\partial x]_1 = [\partial y/\partial x]_2$, and so $\partial y/\partial x$ is continuous along the string. So in addition to the boundary condition (5.20), we have the condition at the boundary

$$\left(\frac{\partial y_1}{\partial x}\right)_{x=0} = \left(\frac{\partial y_2}{\partial x}\right)_{x=0}. \tag{5.21}$$

Let us substitute the expressions for y_i, y_t and y_r, (5.17), (5.18) and (5.19), into (5.20) putting $x = 0$. We get

$$A_i \exp i\omega_i t + A_r \exp i\omega_r t = A_t \exp i\omega_t t. \tag{5.22}$$

If we differentiate (5.17) with respect to x we get

$$\frac{\partial y_i}{\partial x} = -iA_i k_i \exp i(\omega_i t - k_i x),$$

with almost identical expressions for $\partial y_t / \partial x$ and $\partial y_r / \partial x$; if we now substitute these into (5.21) putting $x = 0$, we get

$$-A_i k_i \exp i\omega_i t + A_r k_r \exp i\omega_r t = -A_t k_t \exp i\omega_t t. \tag{5.23}$$

Now equation (5.22) involves real and imaginary quantities, and it has to hold for all values of the time t. This can happen only if the exponential terms cancel out, which means that $\omega_i = \omega_r = \omega_t$. (We are here assuming that A_i, A_r and A_t are all finite.) This is a result that could be predicted physically, for it is clearly impossible for the frequencies of the transmitted and reflected waves to be different from that of the incident wave. We can now replace ω_i, ω_r and ω_t by the single quantity ω.

With this substitution, (5.22) becomes

$$A_i + A_r = A_t \tag{5.24}$$

and (5.23) becomes

$$A_i k_i - A_r k_r = A_t k_t. \tag{5.25}$$

It will be shown in Chapter 6 that

$$c = \sqrt{\frac{T}{\mu}} = \frac{\omega}{k}.$$

Therefore $k_i = k_r$, that is, the incident and reflected waves have the same circular wavenumber. Also,

$$k_i = \omega \sqrt{\frac{\mu_1}{T}} \quad \text{and} \quad k_t = \omega \sqrt{\frac{\mu_2}{T}} \tag{5.26}$$

We can rewrite (5.25) as

$$A_i k_i - A_r k_i = A_t k_t. \tag{5.27}$$

Solving the simultaneous equations (5.24) and (5.27) for A_t and A_r, we get

$$A_t = A_i \frac{2k_i}{k_i + k_t}, \tag{5.28}$$

$$A_r = A_i \frac{k_i - k_t}{k_i + k_t}. \tag{5.29}$$

We now know the amplitudes of the reflected and transmitted waves in terms of the amplitude of the incident wave. Further, since the quantities on the right-hand sides of (5.28) and (5.29) are all real, A_t and A_r are both real numbers and there is no phase change involved, *except* when $k_t > k_i$, whereupon A_r has the opposite sign to A_i. This indicates that the reflected wave has undergone a phase change of π radians with respect to the incident wave.

Now we can introduce the tension and masses per unit length of the two parts of the string from (5.26). When we do this we get

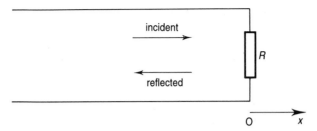

Figure 5.9 Transmission the terminated by a resistance *R*.

$$A_t = \frac{A_i 2\sqrt{\mu_1}}{\sqrt{\mu_1} + \sqrt{\mu_2}}$$

and

$$A_r = A_i \frac{\sqrt{\mu_1} - \sqrt{\mu_2}}{\sqrt{\mu_1} + \sqrt{\mu_2}},$$

so the change of phase on reflection occurs only when $\mu_2 > \mu_1$, that is, when the mass per unit length of the string on the right is greater than that on the left.

This result is of more general applicability. It is always found that when a wave is reflected at a 'denser' medium, a phase change of π occurs, but at a 'less dense' medium, there is no phase change.

Of special interest are the cases where $\mu_2 \gg \mu_1$ and $\mu_2 \ll \mu_1$. If $\mu_2 \gg \mu_1$, the second string is much heavier than the first and may be considered as a rigid boundary. We see that $A_t \approx 0$ and $A_r \approx -A_i$, which is in accord with the result from the previous section that the wave is entirely reflected and inverted.

The case $\mu_2 \ll \mu_1$ corresponds to a very light second string, and the end of the first string at the boundary may be considered 'free' (except that the second string is necessary to provide the tension in the first). Here $A_t \approx 2A_i$ and $A_r \approx A_i$. Again there is reflection, but this time with no phase change.

EXAMPLE 5.1 Reflection of a wave on a transmission line terminated by a resistance *R*

In section 6.8 it will be shown how a transmission line (for example, two parallel wires or a coaxial cable) carries voltage (and current) waves. The speed is $\sqrt{(1/LC)}$, where L and C are respectively the inductance and capacitance per unit length.

Consider a sinusoidal wave travelling along the line (Fig. 5.9). The voltage can be represented by $V = A \exp j(\omega t \mp kx)$, the sign ($\mp$) depending on the direction of travel. Note that we have used j to denote $\sqrt{-1}$ to avoid confusion with i, which denotes current.

The reader will easily verify from (6.41) or (6.42) that

$$\sqrt{\frac{L}{C}} i = \pm A \exp j(\omega t \mp kx)$$

since the wave speed $\omega/k = \sqrt{(1/LC)}$. The quantity $\sqrt{(L/C)}$ is called the *characteristic impedance* of the line, denoted Z_0.

If the incident and reflected waves have amplitudes A and A' respectively, we have therefore

$$V = A \exp \mathrm{j}(\omega t - kx) + A' \exp \mathrm{j}(\omega t + kx)$$

$$Z_0 i = A \exp \mathrm{j}(\omega t - kx) - A' \exp \mathrm{j}(\omega t + kx).$$

Now at the termination $x = 0$, the boundary condition is, from Ohm's law, $V = iR$, and therefore

$$\frac{A + A'}{A - A'} = \frac{R}{Z_0}$$

or

$$\frac{A'}{A} = \frac{R - Z_0}{R + Z_0}.$$

Examination of this result illustrates some important points concerning wave motion, which are applicable in general and not merely to the particular case of waves on a transmission line:

1. If $R < Z_0$, A' and A are of opposite signs, i.e. reflection is accompanied by a change in phase of $180°$. As mentioned earlier, this is the behaviour of any kind of wave which is incident upon a 'denser' medium.

   Likewise, if $R > Z_0$, reflection occurs with no phase change, and this is found to be so for any wave incident upon a 'less dense' medium.

2. If $R = 0$ or $R = \infty$, the amplitude of the reflected wave equals that of the incident wave. This is not surprising since there is no dissipation of power in the resistance in either case; therefore all the incident power is reflected.

3. If $R = Z_0$, there is no reflected wave. The load resistance is said to be 'matched' to the line. Under this condition, maximum power is transferred from the line into the resistance.

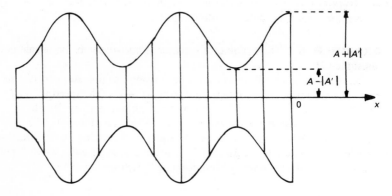

Figure 5.10 Standing wave pattern.

4. The existence of two waves travelling in opposite directions gives rise to a standing wave. But it should be noted that since $|A'|$ is not in general equal to A, we do not in general have points of *no* disturbance, but rather of *minimum* disturbance. Clearly, there are points (Fig. 5.10) where the two waves are in phase, and the voltage has its maximum amplitude $A + |A'|$. Midway between these, they are in antiphase, and the voltage has its minimum amplitude $A - |A'|$. At the termination $x = 0$, V is a maximum if $R > Z_0$ (this case is shown in Fig. 5.10), and a minimum if $R < Z_0$.

An important quantity in electrical engineering is the *voltage standing wave ratio*, which is $(A + |A'|)/(A - |A'|)$. ∎

5.6 Specific acoustic impedance

In the previous section, we referred to the behaviour of waves incident on a boundary with a 'denser' or 'less dense' medium, without at that stage defining clearly what exactly is meant by 'dense'. (It was not intended to mean density in the true sense of the word.) The correct expression for this 'density' is *specific acoustic impedance*, which will now be explained with reference to a longitudinal wave in a solid, or a sound wave in a gas. Waves in these media are treated in detail in Chapter 6, and the reader may need to refer to sections 6.5 and 6.6, but need not study them in detail at this stage.

The *impedance* of a medium is a measure of its 'reluctance' to transmitting a wave. If a given external source provides sinusoidal fluctuations in pressure, the resulting wave will produce a vibration of the particles of the medium. If the medium has a high impedance, the amplitude of vibration and the velocity are both small; similarly, if the impedance is low, the amplitude and velocity are both large.

The specific acoustic impedance Z of the medium is defined as

$$Z = \left| \frac{\text{acoustic pressure}}{\text{particle velocity}} \right| \tag{5.30}$$

for a sinusoidal wave travelling in the medium. Let the disturbance be

$$z = A \exp i(\omega t \mp kx),$$

the upper sign holding for a left-to-right wave, and the lower for right-to-left. It will be shown (6.24) that the acoustic pressure is

$$p = -K \frac{\partial z}{\partial x}$$

where K is the bulk modulus of the medium. Hence

$$p = \pm ikKA \exp i(\omega t \mp kx) \tag{5.31}$$

The velocity $v = \partial z / \partial t$ is

$$v = i\omega A \exp i(\omega t \mp kx)$$

From (5.30) then,

$$Z = \left| \frac{p}{v} \right| = \frac{kK}{\omega} = \frac{K}{c} \tag{5.32}$$

since $\omega = ck$. Note that the factor i has cancelled out, and that Z is real. Physically this means that the pressure and velocity fluctuations are *in phase* (or antiphase). Finally, since $c = \sqrt{(K/\rho)}$ from (6.28), we find for the specific acoustic impedance

$$\boxed{Z = \rho c.} \tag{5.33}$$

The following are values for Z in units of kg m^{-2} s^{-1}: air 400, water 1.45×10^6 and steel 3.9×10^7.

It is a simple matter to show that (5.32) also applies to a longitudinal wave in a rod. This is left as an exercise for the reader, who should refer to section 6.5. (Here, $p = F/A$, where A is the cross-sectional area.)

The importance of specific acoustic impedance can be seen by considering the problem of a sinusoidal wave in a medium, incident upon a boundary with a second medium (refer to Fig. 5.7, and read 'tube' or 'rod' for 'elastic string'). As was seen in section 5.5.2, ω must have the same value throughout, and k is the same for the incident and reflected waves (k_1, say). The equations for the incident, transmitted and reflected waves can be written respectively, since $\omega/k = K/Z$ from (5.32), as

$$z_i = A_i \exp \mathrm{i}(\omega t - k_1 x) = A_i \exp \mathrm{i}\omega(t - Z_1 x/K_1), \tag{5.34}$$

$$z_t = A_t \exp \mathrm{i}(\omega t - k_2 x) = A_t \exp \mathrm{i}\omega(t - Z_2 x/K_2), \tag{5.35}$$

$$z_r = A_r \exp \mathrm{i}(\omega t + k_1 x) = A_r \exp \mathrm{i}\omega(t + Z_1 x/K_1). \tag{5.36}$$

The first boundary condition is $[z_i + z_r]_{x=0} = [z_t]_{x=0}$ since the displacement in both media must be the same at the boundary. Also, the acoustic pressure p is continuous (this point is expanded in section 9.2), which gives the second boundary condition

$$\left[K_1 \frac{\partial z_i}{\partial x} + K_1 \frac{\partial z_r}{\partial x} \right]_{x=0} = \left[K_2 \frac{\partial z_t}{\partial x} \right]_{x=0}.$$

Substitution of these boundary conditions into (5.34), (5.35) and (5.36) gives

$$A_i + A_r = A_t \tag{5.37a}$$

and

$$-\mathrm{i}\omega Z_1 A_i + \mathrm{i}\omega Z_1 A_r = -\mathrm{i}\omega Z_2 A_t,$$

so

$$Z_1(A_i - A_r) = Z_2 A_t \tag{5.37b}$$

The simultaneous equations (5.37a) and (5.37b) are easily solved, yielding

$$A_t = A_i \frac{2Z_1}{Z_1 + Z_2}, \quad A_r = A_i \frac{Z_1 - Z_2}{Z_1 + Z_2} \tag{5.38}$$

The resemblance between these and the corresponding equations for the two strings joined together and the transmission line is quite striking, and the points mentioned previously apply here too, as we now show.

If $Z_2 > Z_1$, there is a phase change of π on reflection, but if $Z_2 < Z_1$, there is no phase change. If $Z_2 = Z_1$, $A_r = 0$, so there is no reflected wave. All the incident energy is therefore transmitted into the second medium. This is called *impedance matching*, and is of great importance where it is vital to transmit as much power as possible from one medium to another.

For example, a television receiving aerial is designed to have an output impedance of 50 Ω. This should be connected to the receiving set by a transmission line (coaxial cable) with a characteristic impedance Z_0 of 50 Ω to ensure that the power generated in the aerial is all transmitted to the cable.

An example in acoustics concerns the design of horns used on old-fashioned gramophones. The purpose of the horn is to provide impedance matching between the needle and the air in the room. Further discussion is beyond the scope of this book.

5.7 Waveguides

5.7.1 Guided waves between parallel plates

Now that we have covered standing waves, we are in a position to consider another important example – namely the waveguide. The most usual meaning of the term 'waveguide' is a metallic tube having rectangular cross-section of the order of a few cm^2, through which electromagnetic waves in the microwave region flow. There are, however, other familiar examples of guided waves. Large houses built before the invention of the telephone were frequently equipped with 'speaking tubes'. A tube such as a hose-pipe provides a very efficient guide for acoustic waves and enables conversations to be conducted over a distance of many metres. The medical stethoscope is another example of an acoustic waveguide.

In the present context, we use the term 'waveguide' to mean any kind of enclosure or container within which a wave may flow. The simplest example to treat mathematically is that of a pair of infinite parallel planes, the wave being contained entirely between them. Though hardly a convincing physical example, it does bring out the essential physics of waveguides, and the results we shall obtain are applicable to other forms of waveguide.

The waveguide is shown in Fig. 5.11 and consists of two infinite parallel planes a distance a apart; the upper plane is along $y = 0$ and the lower along $y = -a$. A monochromatic plane wave of circular wavevector $\mathbf{k}$ is incident on the upper plane as shown. The axes are chosen so that $\mathbf{k}$ lies in the Oxy plane and therefore has no z component. The wave may be of any kind in which the disturbance can be described

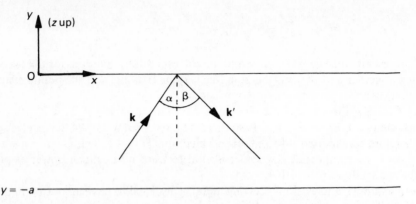

Figure 5.11 A waveguide consisting of two infinite parallel reflecting planes at $y = 0$ and $y = -a$

by a scalar quantity ϕ, for example a sound wave. We represent this incident wave by the exponential form of (4.36):

$$\phi_1 = A_1 \exp \mathrm{i}(\mathbf{k.r} - \omega t). \tag{5.39}$$

The wave produced by reflection at the plane at $y = 0$ can likewise be represented by

$$\phi_2 = A_2 \exp \mathrm{i}(\mathbf{k'.r} - \omega t). \tag{5.40}$$

We make no supposition about the direction of $\mathbf{k'}$ except that it has to be 'generally downward' as shown in Fig. 5.11. The resultant disturbance within the space between the two planes is given by the principle of superposition (section 5.2); we simply add (5.39) and (5.40) to obtain

$$\Phi = \phi_1 + \phi_2 = A_1 \exp \mathrm{i}(\mathbf{k.r} - \omega t) + A_2 \exp \mathrm{i}(\mathbf{k'.r} - \omega t). \tag{5.41}$$

If we express $\mathbf{k}$ and $\mathbf{k'}$ in terms of their components as in section 4.4.2, (5.41) becomes

$$\Phi = A_1 \exp \mathrm{i}(k_x x + k_y y - \omega t) + A_2 \exp \mathrm{i}(k'_x x + k'_y y + k'_z z - \omega t). \tag{5.42}$$

The plane at $y = 0$ imposes the boundary condition $\Phi = 0$ when $y = 0$ for all values of t, in particular $t = 0$. Applying these in (5.42) we have

$$A_1 \exp \mathrm{i}(k_x x) + A_2 \exp \mathrm{i}(k'_x x + k'_z z) = 0. \tag{5.43}$$

Now (5.43) must be true for all values of z, which can be so only if

$$k'_z = 0. \tag{5.44}$$

We see at once that $\mathbf{k'}$ must lie in the Oxy plane.

Again, since (5.43) must be true for all values of x, we must have

$$k_x = k'_x \tag{5.45}$$

therefore

$$A_1 = -A_2. \tag{5.46}$$

Now $|\mathbf{k}| = |\mathbf{k}'|$ since the wavelength, and hence $|\mathbf{k}|$, are determined solely by the medium for a wave of given frequency. It follows from (5.44) and (5.45) that

$$k_y = -k_y'. \tag{5.46a}$$

(We reject the solution $k_y = k_y'$ as being inconsistent with the physics of the problem – see Fig. 5.11.)

Thus the component of the circular wavevector $\mathbf{k}$ parallel to the reflection plane is unaltered on reflection, but that perpendicular to the plane is reversed. This requires that both wave normals make the same angle with the plane normal, so that $\alpha = \beta$ in Fig. 5.11. This is the law of reflection, which is treated further in Chapter 10.

Substituting (5.44), (5.45), (5.46) and (5.46a) into (5.42) and writing $A_1 = A$, we have

$$\Phi = A \exp i(k_x x + k_y y - \omega t) - A \exp i(k_x x - k_y y - \omega t)$$

$$= 2A \mathrm{i} \sin k_y y \exp i(k_x x - \omega t). \tag{5.47}$$

So (5.47) describes in complex exponential form the resultant wave in the space between the two planes at $y = 0$ and $y = -a$; there is no need to take the real part of (5.47) for the interpretation we now make.

The presence of the plane at $y = -a$ imposes a further boundary condition on (5.47); when $y = -a$, $\Phi = 0$ so $\sin(k_y a) = 0$ no matter what value x has. Thus $k_y a = \pm n\pi$ or

$$k_y = \pm \frac{n\pi}{a} \quad \text{where } n = 1, 2, 3 \ldots. \tag{5.48}$$

The only allowed values of k_y are those given by (5.48).

We can now give a full physical interpretation of the wave given by (5.47). We recognise $\exp i(k_x x - \omega t)$ as describing a travelling wave propagating in the x-direction. The amplitude of the disturbance propagated is given by $2A \sin(n\pi y/a)$ and therefore varies in the y-direction. Sketches showing the form of this variation are given in Fig. 5.12, and these follow rather obviously from the above expression.

Thus we have a wave which is a *travelling* wave in the x-direction but shows the characteristics of a *standing* wave in the y-direction. An interesting feature of this wave is that not all frequencies may be propagated, for

$$k^2 = k_x^2 + k_y^2,$$

so, by (5.48),

$$\frac{\omega^2}{c^2} = k_x^2 + \frac{n^2 \pi^2}{a^2} \tag{5.49}$$

where $c = \omega/k$ is the wave speed and hence

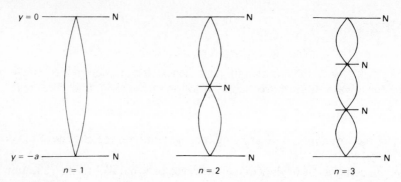

Figure 5.12   Variation of sin($n\pi y/a$) with y for $n=1$, 2 and 3. The nodal planes (i.e. the planes of zero disturbance at all times) are indicated by N.

$$k_x = \sqrt{\left(\frac{\omega^2}{c^2} - \frac{n^2\pi^2}{a^2}\right)}$$

Thus only waves of frequency greater than $c/2a$ may be guided.

A second interesting feature is that the wave travels with a speed greater than that of the original incident wave. The speed of the wave of (5.47) is $c' = \omega/k_x$; thus $c' > c$ since $k_x < k$, k_x being a component of **k**.

This is at first a somewhat puzzling result when applied to electromagnetic waves, since it would seem to imply a speed greater than that of light. But if we may anticipate a result from section 6.10, namely that the speed at which energy is propagated is given by the group velocity

$$c_g = d\omega/dk,\tag{6.52}$$

we see by differentiating (5.49) with respect to k_x that

$$\frac{2\omega}{c^2}\frac{d\omega}{dk_x} = 2k_x,$$

so

$$c_g = \frac{d\omega}{dk_x} = \frac{c^2 k_x}{\omega},$$

hence

$$c'c_g = c^2.\tag{5.50}$$

Thus while the phase velocity c' is greater than c, it follows from (5.50) that the group velocity c_g is less than c.

The above treatment can be adapted fairly readily for enclosed waveguides of rectangular cross-section, and with rather more difficulty for guides of circular cross-section. What is found in both cases is a progressive wave along the guide but with standing-wave characteristics in transverse directions. But the main essentials of waveguides are brought out by the simplified treatment we have given.

5.7.2 Optical fibres

The principles developed in the previous section have their most interesting and technologically important application in the field of telecommunications. Light waves generated from laser diodes are guided along cylindrical fibres of pure glass of diameter half a millimetre or under. The fibre is coated with a layer having a refractive index about 10% lower than the glass itself so that total internal reflection takes place. The power losses due to leakage and absorption are so low that the wave can travel several kilometres before amplification is needed.

The most significant recent advance is the development of optical amplification which enables the signal to be boosted directly and *in situ*. This is achieved by mixing the weak incoming light with light of slightly lower wavelength in a section of guide doped with rare-earth ions. Previously, amplification required the optical signal to be converted first into an electrical signal which was amplified electronically and then converted back – clearly a much more complicated and slower process. Optical fibres are now used in sub-ocean cables where the signal is first converted to a series of binary pulses for reliable transmission. Binary information can now be transmitted along such fibres at a rate of 5×10^9 binary 'bits' per second, i.e. 5 Gbit s^{-1}. This means that a succession of pulses, each of amplitude either zero or unity, and of duration $1/(5 \times 10^9) = 2 \times 10^{-10}$ s, can be passed along the fibre.

It is interesting to estimate how many telephone conversations can be carried by a single fibre. It is not necessary to transmit analogue speech over the whole of the audible range of 0–20 kHz. Perfectly intelligible speech may be obtained by transmitting only those frequencies in the range 300–3400 Hz and this is in fact known as the telephone range. In practice the voice signal is sampled at the rate of 8000 times per second (8 kHz), rather more than double the telephone range, and the amplitude of the wave at each sample point is measured on a scale of 1–8. (This is done on a logarithmic, or decibel, scale for reasons which are made clear in section 9.5.) Thus the transmission of a single telephone conversation requires $8000 \times 8 = 64\,000$ bit s^{-1}, which means that a fibre of capability 5 Gbit s^{-1} can carry $5 \times 10^9/64\,000$ or 78\,000 messages simultaneously.

If high-quality sound is to be transmitted, the number of separate programmes the fibre can carry is much reduced. For example, the so-called CD (compact disc) Quality standard requires that the full audio range is covered, which results in the audio signal being sampled 44.1×10^3 times per second. Each sampled amplitude is now placed on a logarithmic scale of 1–16 which means that approximately 7×10^5 bit s^{-1} are required for a single audio channel. Thus around 7000 high-quality sound programmes can be transmitted simultaneously. It is easy to understand therefore the enormous current interest and activity in the field of fibre-optic communication.

Problems

1. By plotting rough sketches, investigate the behaviour of a disturbance of the same shape as that given in Fig. 5.5, which is reflected at a boundary *without* inversion. [(5.29), with $k_i > k_t$, shows that this can occur.]

2. Two stretched strings are joined, the masses per unit length being 0.01 kg m^{-1} and 0.04 kg m^{-1}. A sinusoidal wave of amplitude A is propagated along the first towards the join. Deduce the amplitude of (i) the transmitted wave and (ii) the reflected wave. Is there a phase change of the reflected wave?

    Sketch the shape of the strings at the moment of maximum displacement of the join.

3. Using the idea of acoustic impedance, suggest a reason why a tuning fork vibrating in air transmits very little power to the air, but sounds loud if its stem is held rigidly against a large solid bench. (See also section 9.4.)

4. A coaxial cable has a characteristic impedance of 80 Ω. Assuming that waves travel along it with a speed 1.5×10^8 m s^{-1}, calculate the inductance and capacitance per unit length.

5. A transmission line with characteristic impedance 50 Ω is terminated by a resistance 10 Ω. Is there a phase inversion on reflection? Determine the fraction of the incident power absorbed and reflected. Deduce the voltage standing wave ratio.

6 Waves in physical media

6.1 Introduction

In Chapter 4 we derived the partial differential equation governing wave propagation in one dimension,

$$\frac{\partial^2 y}{\partial x^2} = \frac{1}{c^2} \frac{\partial^2 y}{\partial t^2},$$

[4.12]

from purely geometrical considerations. We assumed a wave profile was being transmitted without change of shape and with constant velocity, but gave no physical justification for these assumptions. Let us therefore examine some instances of wave propagation which will lead to equations of the type (4.12).

6.2 Transverse waves in an infinitely long, stretched elastic string

Suppose an infinitely long elastic string is initially at rest (in equilibrium) and lies along the Ox-axis of coordinates. We will denote the transverse disturbance due to the passage of the wave by y. Let the mass per unit length of string be μ and let the tension in the string be T.

We will make the following assumptions:

1. The value of y is very small compared with any wavelength with which we are concerned, and the disturbed string makes small angles with the x-axis, i.e. $\partial y / \partial x \ll 1$.
2. There is no motion other than in the y-direction.
3. The tension in the string is unaltered by the passage of the wave.
4. The effects of gravity can be ignored, i.e. the weight of the string is left out of our considerations.

We now obtain the partial differential equation governing the propagation of transverse waves along the string, by applying Newton's second law of motion to a short element of the string at an instant during the passage of the wave. Consider

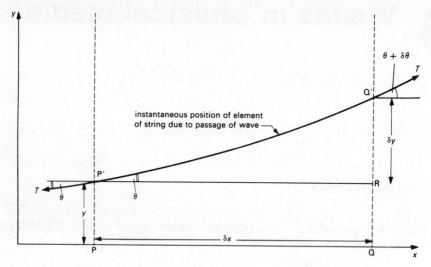

Figure 6.1 Element of string displaced by passage of wave.

the motion of the element of length δx whose equilibrium position is PQ (Fig. 6.1). At an instant during the passage of the wave, the element is displaced to the instantaneous position P′Q′.

The angles made by the string with the Ox-axis at the points P′ and Q′ are θ and $\theta + \delta\theta$ as shown in Fig. 6.1; the only forces acting on the element are the tensions exerted by the adjoining sections of strings at P′ and Q′. The net force acting in the Oy-direction is therefore

$$T \sin(\theta + \delta\theta) - T \sin\theta \approx T(\theta + \delta\theta) - T\theta = T\delta\theta = T\frac{\partial\theta}{\partial x}\delta x \qquad (6.1)$$

since θ and $\theta + d\theta$ are small, by assumption (1).

Also, since θ is small,

$$\tan\theta \cong \theta = \frac{\delta y}{\delta x}\ (\triangle P'Q'R\ \text{of Fig. 6.1}). \qquad (6.2)$$

Therefore, in the limit as $\delta x \to 0$

$$\theta = \frac{\partial y}{\partial x}, \quad \text{so} \quad \frac{\partial\theta}{\partial x} = \frac{\partial^2 y}{\partial x^2},$$

and the net force on the element, by (6.1), is

$$T\frac{\partial^2 y}{\partial x^2}\delta x. \qquad (6.3)$$

From Newton's second law of motion, this equals the product of mass and acceleration for the element

$$(\mu \, \delta x) \frac{\partial^2 y}{\partial t^2},$$

hence, finally

$$\frac{\partial^2 y}{\partial x^2} = \frac{1}{T/\mu} \frac{\partial^2 y}{\partial t^2}. \tag{6.4}$$

This equation is in the same form as equation (4.12), so we see at once, subject to all the assumptions we have made, that transverse waves of any profile can be propagated along a stretched elastic string, and that they will travel with the unique speed $\sqrt{(T/\mu)}$. We can alter this speed only by changing the tension or the mass per unit length of the string (or both).

6.3 Energy in a travelling wave

We now consider the propagation of energy in a wave, taking transverse waves on a string as our example.

In the absence of a wave, the kinetic energy is zero since no particle in the string has any velocity, and, since we can choose the zero of potential energy arbitrarily, we will assume that the potential energy is also zero. When a travelling wave of transverse displacement $y(x, t)$ is being propagated along the string, the latter is stretched locally, thus acquiring potential energy. Figure 6.2 shows an infinitesimal length of the string, from which we see that a section of the string, which in the absence of a wave is of length dx, has now a length ds. The section has thus been stretched by an amount $ds - dx$. The constant force exerted during this stretching is T, the tension in the string, and thus the potential energy dE_P acquired by the length is

$$dE_P = T(ds - dx),$$

which, by reference to Fig. 6.2, becomes

$$dE_P = T[(dx^2 + dy^2)^{1/2} - dx]$$

$$= T \, dx \left\{ \left[1 + \left(\frac{\partial y}{\partial x} \right)^2 \right]^{1/2} - 1 \right\}.$$

By the binomial theorem this finally becomes, approximately,

$$dE_P = \tfrac{1}{2} T \left(\frac{\partial y}{\partial x} \right)^2 dx. \tag{6.5}$$

The kinetic energy dE_K of the element is

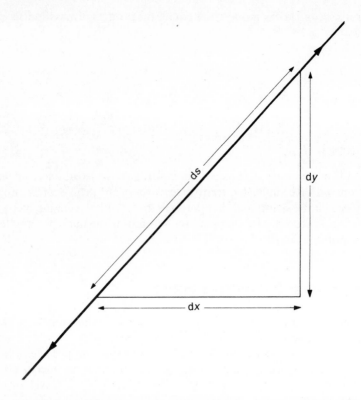

Figure 6.2 A section of stretched string while a transverse wave (upward displacement) is being propagated horizontally, showing the relationship between ds, dy and dx.

$$dE_K = \tfrac{1}{2}(\text{mass of element}) \times (\text{velocity of element})^2$$

$$= \tfrac{1}{2}\mu \, dx\left(\frac{\partial y}{\partial t}\right)^2. \tag{6.6}$$

If we denote the total energy of the element by dE, we see from (6.5) and (6.6) that

$$dE = \frac{dx}{2}\left[\mu\left(\frac{\partial y}{\partial t}\right)^2 + T\left(\frac{\partial y}{\partial x}\right)^2\right].$$

We now define the *energy density*, $D(x, t)$ as the energy per unit length. Thus

$$D(x, t) = \frac{1}{2}\left[\mu\left(\frac{\partial y}{\partial t}\right)^2 + T\left(\frac{\partial y}{\partial x}\right)^2\right]. \tag{6.7}$$

Let us calculate D for the harmonic wave represented by

$$y = A \cos(kx - \omega t). \tag{6.8}$$

If we differentiate (6.8) partially and substitute into (6.7) we obtain

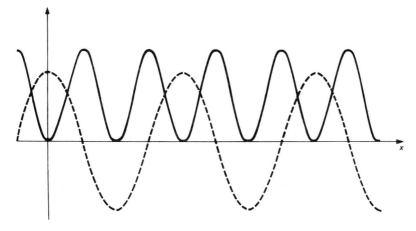

Figure 6.3 The dotted line represents the profile of the displacement wave (6.8), and the solid line the corresponding energy-density wave (6.9) for time $t = 0$. The scales of the ordinates are arbitrary.

$$D(x, t) = \frac{A^2}{2} \sin^2(kx - \omega t)[\mu\omega^2 + Tk^2]$$

$$= \frac{\mu A^2}{2} \sin^2(kx - \omega t)\left[\omega^2 + \frac{T}{\mu} k^2\right].$$

Since $T/\mu = c^2$ the contents of the square bracket become $[\omega^2 + c^2k^2]$, which, since $\omega^2 = c^2k^2$, finally become $2\omega^2$. Notice that since the two bracketed terms are equal, *the kinetic energy and potential energy contribute equally* to the total energy in the wave. Thus

$$D(x, t) = \mu A^2 \omega^2 \sin^2(kx - \omega t). \tag{6.9}$$

So we see that the energy in the wave is propagated as the function of $(kx - \omega t)$ described in the right-hand side of (6.9); it is thus propagated with the speed $\omega/k = c$ of the displacement wave.

Actually, the circular frequency ω and circular wavenumber k of the displacement wave are not those of the energy-density wave. Since

$$\sin^2 \theta = \tfrac{1}{2}(1 - \cos 2\theta),$$

it follows that

$$D(x, t) = \frac{\mu A^2}{2} \omega^2[1 - \cos(2kx - 2\omega t)]. \tag{6.10}$$

Thus the energy-density wave propagates with a frequency twice that of the displacement wave and a wavelength of half that of the displacement wave, but, as mentioned above, the *velocity* is the same as that of the displacement wave. The spatial parts of the energy-density and displacement waves are shown in Fig. 6.3.

The energy in length dx is $D(x, t) \, dx$, and therefore the energy in one wavelength is $\int_0^\lambda D(x, t) \, dx$, which, from (6.10), becomes

$$\int_0^\lambda \frac{\mu A^2}{2} \, \omega^2 [1 - \cos(2kx - 2\omega t)] \, dx.$$

The integral of the cosine term is zero; and the energy contained in a wavelength is thus

$$\int_0^\lambda \frac{\mu A^2}{2} \, \omega^2 \, dx = \frac{\mu A^2 \omega^2 \lambda}{2}.$$

Now this energy takes a time equal to the period λ/c to pass a given point. Therefore the energy flowing per unit time (the power P) is given by

$$P = \frac{\mu A^2 \omega^2 c}{2}. \tag{6.11}$$

We see, therefore, that the power transmitted by the wave is proportional to the square of the amplitude and the square of the frequency. A similar result is obtained for many other types of harmonic wave, including sound waves. That light intensity is proportional to the square of the amplitude of the disturbance is an analogous result which will be used in Chapter 10.

6.4 Transmission and reflection coefficients

Consider again a transverse wave in a string, which is fixed to a second string with a different value of μ. This was analysed in section 5.5.2 and led to equations (5.28) and (5.29) for the transmitted and reflected wave amplitudes. We now find the power in each of these waves.

Now if F is the tension in the strings, $c = \sqrt{(F/\mu)}$ so $\mu = F/c^2$. Hence equation (6.11) can be written

$$P = \frac{\mu A^2 \omega^2 c}{2} = \frac{F A^2 \omega^2}{2c} = \frac{F A^2 \omega k}{2} \tag{6.12}$$

The tension F and circular frequency ω are the same in both strings, so the power P is proportional to $A^2 k$. The fractional power transmitted (the transmitted power divided by the incident power) is therefore, from (5.28),

$$T = \frac{A_t^2 k_2}{A_i^2 k_1} = \frac{4k_1 k_2}{(k_1 + k_2)^2} \tag{6.13}$$

which is termed the *transmission coefficient*.

Likewise, the *reflection coefficient* R is the fractional power reflected (the reflected power divided by the incident power) and is, from (5.29),

$$R = \frac{A_r^2 k_1}{A_i^2 k_1} = \frac{(k_1 - k_2)^2}{(k_1 + k_2)^2} \tag{6.14}$$

The reader will quickly verify that $R + T = 1$ as must clearly be the case, since energy is conserved.

The above discussion can be extended to the case of waves at interfaces or boundaries in other media. In particular, for sound waves impinging on a different medium, or longitudinal waves in two joined rods, it can be shown by an analysis similar to that given in section 6.3 that the power per unit cross-sectional area, known as the *intensity*, is

$$I = \frac{\rho A^2 \omega^2 c}{2}$$

where A is again the amplitude. This is very similar to (6.11). Introducing $Z = \rho c$ (5.33), we have

$$I = \frac{A^2 \omega^2 Z}{2}$$

Finally, we can write the transmission and reflection coefficients in terms of the specific acoustic impedance as follows, using (5.38)

$$T = \frac{I_t}{I_i} = \frac{A_t^2 Z_2}{A_i^2 Z_1} = \frac{4 Z_1 Z_2}{(Z_1 + Z_2)^2} \tag{6.15}$$

$$\cdot R = \frac{I_r}{I_i} = \frac{A_r^2 Z_1}{A_i^2 Z_1} = \frac{(Z_1 - Z_2)^2}{(Z_1 + Z_2)^2} \tag{6.16}$$

and again, $R + T = 1$.

6.5 Longitudinal waves in a rod

Suppose we have an infinitely long rod (Fig. 6.4), of uniform cross-section of area A made of material of density ρ and Young's modulus E. Consider the small section of length δx between P and Q, shown in Fig. 6.4a, originally at rest. Suppose the rod is set into longitudinal agitation, for example by hitting the rod at its left end. Clearly this will set the adjacent part of the rod into longitudinal motion, and this disturbance will pass along the rod in the form of a longitudinal wave. Suppose that at a given instant the cylinder PQ has become displaced to a new position such that P moves a distance z to P' and Q moves a distance $z + \delta z$ to Q'. The variable z measures the longitudinal displacement of a point due to the passage of the wave. Let the force on the left-hand face now be F and that on the right-hand face $F + \delta F$.

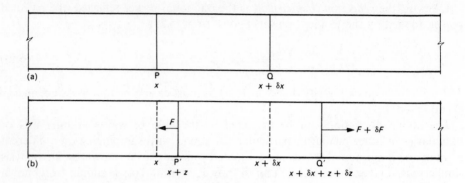

Figure 6.4 Longitudinal waves in a rod: (a) at rest, while in (b) the section PQ is displaced to P′Q′ owing to the passage of the wave.

We derive the differential equation governing the propagation of longitudinal waves in two steps. First we relate the change in dimensions of the cylinder to the force acting; then we equate the rate of change of momentum of the cylinder to the net force acting upon it, by Newton's second law of motion.

Relating change in length to force acting, we have, from the definition of the Young modulus,

$$\frac{F}{A} = E\,\frac{\delta z}{\delta x},\tag{6.17}$$

ignoring δF in comparison with F. Applying Newton's second law of motion we have

$$\delta F = A\rho\,\delta x\,\frac{\partial^2 z}{\partial t^2}.$$

But

$$\delta F = \frac{\partial F}{\partial x}\,\delta x$$

$$= AE\,\frac{\partial^2 z}{\partial x^2}\,\delta x \quad \text{by (6.17)}.$$

Hence

$$\frac{\partial^2 z}{\partial x^2} = \frac{1}{E/\rho}\frac{\partial^2 z}{\partial t^2},\tag{6.18}$$

which is the wave equation once again. The velocity for longitudinal waves in a rod is thus

$$c = \sqrt{\frac{E}{\rho}}$$

(6.19)

This gives velocities of the order of 5000 m s^{-1} for typical values of E and ρ of common metals.

6.6 Longitudinal waves in a fluid (liquid or gas)

The reasoning of the previous section can be applied with modification to the case of longitudinal waves in a fluid. Suppose we have an infinitely long hollow tube (Fig. 6.5), of cross-section of area A, containing a fluid (liquid or gas). Suppose that originally the fluid is at rest, its density is ρ and the pressure is P_0. Consider the small cylinder of fluid of length δx between R and S, shown in Fig. 6.5a, originally at rest with an equal pressure P_0 exerted at both ends by the surrounding fluid. Suppose again that the fluid is set into longitudinal motion, for example by inserting a piston in the tube somewhere to the left of R and causing it to oscillate axially. Suppose that at a given instant the cylinder RS has become displaced to R′S′. Let the pressure on the left-hand face now be P and that on the right-hand face $P + \delta P$.

The bulk modulus K of a material is a measure of the pressure increase required to change its volume by a given amount. It is defined formally as

$$K = \frac{\text{stress}}{\text{strain}} = \frac{\text{increase in applied pressure}}{\text{fractional change in volume}}.$$

(6.20)

Proceeding to the limit of vanishingly small change in volume, we obtain the differential expression

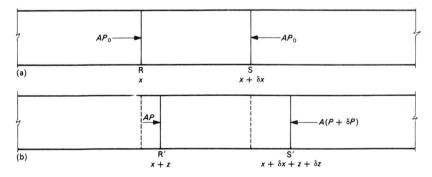

Figure 6.5 Waves in a fluid. In (a) the fluid is at rest in equilibrium, while in (b) the passing wave has displaced the cylinder RS to R′S′.

$$K = -V \frac{\mathrm{d}P}{\mathrm{d}V}. \tag{6.21}$$

The minus sign appears because an increase in pressure produces a decrease in volume.

Applying the definition (6.20) to changes that have occurred to the cylinder RS between Fig. 6.5a and 6.5b, gives

$$K = -\frac{(P - P_0)}{A\delta z / A\delta x}, \tag{6.22}$$

and defining the *acoustic pressure p* as $P - P_0$ gives

$$p = -K \frac{\delta z}{\delta x} \tag{6.23}$$

or, in the limit as $\delta x \to 0$

$$p = -K \frac{\partial z}{\partial x}. \tag{6.24}$$

Here δP has been ignored in comparison with p.

Equating the net force on the displaced cylinder R'S' to the product of mass and acceleration (Newton's second law) gives

$$PA - (P + \delta P)A = (A\rho\delta x) \frac{\partial^2 z}{\partial t^2}$$

or

$$-\delta P = \rho\delta x \frac{\partial^2 z}{\partial t^2}. \tag{6.25}$$

Here we cannot ignore δP, for it is this (small) excess pressure that produces the motion of the cylinder of fluid.

From (6.23) $\delta p = \delta P$, since P_0 is constant, so (6.25) becomes

$$-\delta p = \rho\delta x \frac{\partial^2 z}{\partial t^2}$$

or

$$-\frac{\partial p}{\partial x} \delta x = \rho\delta x \frac{\partial^2 z}{\partial t^2}. \tag{6.26}$$

But, from (6.24),

$$\frac{\partial p}{\partial x} = -K \frac{\partial^2 z}{\partial x^2};$$

substituting this into (6.26) and tidying up gives

$$\frac{\partial^2 z}{\partial x^2} = \frac{1}{K/\rho} \frac{\partial^2 z}{\partial t^2}.$$

(6.27)

Thus longitudinal waves may be propagated in a fluid, and have a speed c given by

$$\boxed{c = \sqrt{\frac{K}{\rho}}}$$

(6.28)

6.7 Pressure waves in a gas

Now equation (6.27) involves z which is the displacement of a point in the fluid. It is not therefore particularly useful, since we are not usually interested in how the longitudinal positions of points within the fluid vary because of the wave. On a molecular scale the position of a point in a fluid is a rather meaningless concept because, as kinetic theory tells us, the molecules in a liquid or a gas are continuously moving about at very high velocities. What is of great importance is the variation of pressure due to waves in fluids – particularly gases. The partial differential equation governing pressure waves is obtained from equation (6.24). If we partially differentiate this twice with respect to time we get

$$\frac{\partial^2 p}{\partial t^2} = -K \frac{\partial^2}{\partial t^2} \left(\frac{\partial z}{\partial x} \right).$$

When we carry out the differentiation of the right-hand side, we obtain the so-called mixed derivative, which is written in the form

$$\frac{\partial^2}{\partial t^2} \left(\frac{\partial z}{\partial x} \right) = \frac{\partial^3 z}{\partial t^2 \, \partial x}.$$

(6.29)

Note that the order in which the terms appear in the denominator of the right-hand side of (6.29) shows the order in which differentiation takes place. Thus

$$\frac{\partial^2 p}{\partial t^2} = -K \frac{\partial^3 z}{\partial t^2 \, \partial x}.$$

(6.30)

Differentiating (6.24) partially with respect to x gives

$$\frac{\partial p}{\partial x} = -K \frac{\partial^2 z}{\partial x^2}.$$

(6.31)

Eliminating $\partial^2 z/\partial x^2$ between (6.27) and (6.31) gives

$$\frac{\partial p}{\partial x} = -\rho \frac{\partial^2 z}{\partial t^2},$$

and partially differentiating this with respect to x gives

$$\frac{\partial^2 p}{\partial x^2} = -\rho \frac{\partial}{\partial x}\left(\frac{\partial^2 z}{\partial t^2}\right) = -\rho \frac{\partial^3 z}{\partial x \partial t^2}. \tag{6.32}$$

It turns out that the order of differentiation in mixed derivatives such as $\partial^3 z/\partial x \partial t^2$ does not matter in the case of nearly all functions met in physics, so we may eliminate this quantity between (6.30) and (6.32) to obtain

$$\frac{\partial^2 p}{\partial x^2} = \frac{1}{K/\rho}\frac{\partial^2 p}{\partial t^2}, \tag{6.33}$$

so the pressure waves have the same speed $\sqrt{(K/\rho)}$ as the displacement waves. Note that p in (6.33) is not the actual pressure, but the *change* from ambient pressure, i.e. the *acoustic pressure* produced by the wave.

The pressure waves in a gas of greatest interest here are sound waves. It can be shown that changes in local conditions produced by the passage of a sound wave take place *adiabatically*; that is, no heat energy enters or leaves any given small region of the gas during the half period between, say, a compression and the following expansion. A popular, but wrong, explanation of this is that sound frequencies are so high that there is not time for the heat energy to get away. The correct explanation of why sound waves are adiabatic is somewhat involved and will not be given here, but it forms the basis for problem 8 at the end of this chapter.

When the state of a fixed mass of ideal gas changes adiabatically, it is easily shown from elementary thermodynamics that

$$PV^\gamma = \text{constant} \tag{6.34}$$

where P is the pressure, V the volume and γ the ratio of the principal heat capacities (C_p/C_v). Taking logs of (6.34) we have

$$\ln P + \gamma \ln V = 0. \tag{6.35}$$

Differentiation and tidying up produces

$$V\frac{dP}{dV} = -\gamma P, \tag{6.36}$$

while substituting for $V\,dP/dV$ from (6.21) gives

$$K_A = \gamma P. \tag{6.37}$$

The suffix A in K_A shows that this is the bulk modulus for adiabatic conditions. Substituting this value for K_A into (6.28) gives the speed c for sound in a gas as

$$c = \sqrt{\frac{\gamma P}{\rho}}. \tag{6.38}$$

Still assuming an ideal gas, which obeys

$$PV_m = RT \tag{6.39}$$

where R is the molar gas constant, T the thermodynamic temperature, and V_m the volume of 1 mole, we may substitute this together with $V_m = M/\rho$, where M is the molar mass, into (6.38) to obtain

$$c = \sqrt{\frac{\gamma R T}{M}}.$$

(6.40)

This tells us that the speed of sound in a gas is independent of the pressure (at constant temperature), and increases with the square root of the absolute temperature; and also that the speed is greatest in gases of low molar mass, since $c \propto M^{-1/2}$ for fixed T.

EXAMPLE 6.1 To calculate the speed of sound in air at 15°C
It will be appreciated that the speed of sound is different in different gases because both M and γ depend on the gas. Since air is a mixture of 80% nitrogen and 20% oxygen, for which the molar masses are 28 and 32 g respectively, M can be taken as the weighted mean $(0.8 \times 28 + 0.2 \times 32) \times 10^{-3} = 28.8 \times 10^{-3}$ kg mol^{-1}. We ignore the presence of other gases such as carbon dioxide and water vapour.

Now γ depends on the number of atoms in a molecule of the gas. For monatomic gases (e.g. He, Ne, A), $\gamma = 5/3 = 1.667$, while for diatomic gases (e.g. H_2, N_2, CO), $\gamma = 7/5 = 1.4$. Since nitrogen (N_2) and oxygen (O_2) are diatomic, γ can be taken with good accuracy as 1.4 for air.

Using the value $R = 8.31$ J K^{-1} mol^{-1}, we can calculate the speed of sound in air at 15°C ($= 288$ K),

$$c = \sqrt{\frac{1.4 \times 8.31 \times 288}{28.8 \times 10^{-3}}} = 341 \text{ m s}^{-1}.$$

(It should be noted that M must be expressed in kg mol^{-1}.)

We repeat the point that the speed of sound in air – or indeed any gas – depends on the temperature, but *not on the pressure*. This might appear to be inconsistent with the equation $c = (\gamma P/\rho)^{1/2}$, until it is realised that if P is increased at constant T, then ρ also increases in proportion. ■

6.8 Current and voltage waves in an ideal transmission line

A transmission line is a system of conductors, usually two, carrying electrical signals. In its simplest form, it consists of a pair of identical, parallel wires, separated by an insulating medium (for example, air or polythene). A simple instance is a length of household twin flex. Another familiar example is a coaxial line, consisting of a solid central wire surrounded by a coaxial cylindrical conductor, as used to connect a television aerial to a set. If one end of the line is connected to an alternating current

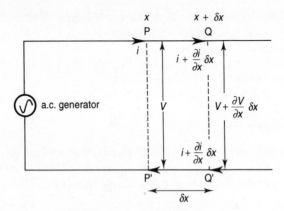

Figure 6.6 Ideal transmission line.

generator (Fig. 6.6), fluctuations in voltage and current are produced, which proceed along the line in the form of an electrical wave.

We will investigate the changes that take place in the small section of the line of length δx between PP′ and QQ′. We assume that the wires have zero resistance, and there is no leakage of current from one to the other. Such a line is said to be 'loss free'.

Now the two wires of the line act as a capacitor. It would appear that the problem of considering line capacitance is formidable, because PQ interacts not only with P′Q′, but with all elements of the lower line. However, it is clear that the electric field at PQ can have no component along PQ; otherwise there would be an infinite current in this element. The electric field is therefore transverse, and all such field lines originating from PQ must end on P′Q′. We are therefore justified in regarding PQ, P′Q′ as an elementary capacitor, and can ignore this capacitance between PQ and other parts of the line.

Let the capacitance per unit length be C. Then the charge on PQ is $+CV\delta x$ (and that on P′Q′ is $-CV\delta x$). Now if V is increasing with time, the charge on PQ is increasing at a rate $C(\partial V/\partial t)\delta x$. This must therefore be equal to $i_P - i_Q = -(\partial i/\partial x)\delta x$, and so

$$\frac{\partial i}{\partial x} = -C\frac{\partial V}{\partial t}. \tag{6.41}$$

We now consider in a similar manner the inductive effect of the line, but must be somewhat cautious. The inductance of a closed circuit can be defined as being (numerically) equal to the magnetic flux threading the circuit when unit current flows in it. But since the current flows along PQ and P′Q′, but not along PP′ and QQ′, PQQ′P′ does not form a closed circuit. Can we talk meaningfully about the inductance of the element PQQ′P′? Well, consider a very long line, terminated at both ends, carrying a *steady* unit current. The flux threading the line is uniformly distributed, and proportional to the length of the line. It is therefore meaningful to define an 'inductance per unit length', L, such that the flux threading the element PQQ′P′ is $Li\delta x$. Like the electric field, the magnetic field is transverse.

Now if i is increasing, the voltages across PP′ and QQ′ are different. The induced voltage in the path PQQ′P′ is $-dV = -(\partial V/\partial x)\delta x$. This equals the rate of change of flux threading the path, and so

$$\frac{\partial V}{\partial x} = -L\frac{\partial i}{\partial t}. \tag{6.42}$$

(As readers familiar with electromagnetism will appreciate, Faraday's law may be applied round a closed *path* as we have done; it does not matter that it is not a complete *circuit*.)

It is easy now to show that voltage and current waves are propagated along the line. For if we partially differentiate (6.41) with respect to t, (6.42) with respect to x, and eliminate the mixed term, we obtain

$$\frac{\partial^2 V}{\partial x^2} = LC\frac{\partial^2 V}{\partial t^2}. \tag{6.43}$$

Likewise, partially differentiating (6.41) with respect to x, and (6.42) with respect to t, leads to

$$\frac{\partial^2 i}{\partial x^2} = LC\frac{\partial^2 i}{\partial t^2}. \tag{6.44}$$

Equations (6.43) and (6.44) represent voltage and current waves, which exist simultaneously, and have speed c given by

$$\boxed{c = \sqrt{\frac{1}{LC}}}. \tag{6.45}$$

Values of L and C for a specimen of ordinary household connecting flex are 7×10^{-7} H m^{-1} and 6×10^{-11} F m^{-1} respectively. The wave velocity is therefore 1.54×10^8 m s^{-1}, which is rather more than half the velocity of light in vacuum (3×10^8 m s^{-1}).

We can calculate the wavelength for 50 Hz alternating current in household flex from the relationship

wave velocity = frequency × wavelength;

this gives λ to be 3090 km. The lengths of the lines over which alternating current is transmitted are very much smaller than this, so that the signals are transmitted practically instantaneously compared with the period of the signal; it is therefore unnecessary to speak of waves in this context, but, when the signal frequency is high, the wave nature can become important.

The wave velocity can be decreased by using cable with a higher value for the product *LC*; a complete analysis of this problem shows that the value of *LC* is determined solely by the nature of the insulation and is independent of the geometry of the conductors.

The time taken for a wave to travel distances of the order of a metre (e.g. about 5×10^{-9} s for coaxial cable) is significant in electronic terms; this is the basis of the so-called delay line. One example of the use of delay line is in the trigger mechanism of good-quality cathode-ray oscilloscopes. When randomly occurring pulses are to be displayed, the leading edge of the pulse has to be used to trigger the sweep mechanism; that is, to start the spot off on its journey across the screen. The incoming pulse amplitude is split into two; the first part is used to trigger the sweep, whilst the second part is passed along a delay line timed so that its leading edge arrives at the plates immediately after the spot has begun to move. In this way, the whole of the pulse waveform is displayed upon the screen.

In this chapter we have so far seen how the equation

$$\frac{\partial^2 y}{\partial x^2} = \frac{1}{c^2} \frac{\partial^2 y}{\partial t^2}$$

arises in a variety of physical situations. It is by no means the only wave equation; any partial differential equation relating a displacement in a medium to the spatial coordinates and time is a wave equation. But the above equation is the simplest general description of the essentials of wave motion. It is known as the non-dispersive wave equation.

6.9 The Doppler effect

Consider a source of sinusoidal waves and an observer remote from the source. If the source is moving, the observer receives a wave of frequency different from that generated by the source. If, on the other hand, the observer is moving, the source being stationary, the observer again receives a wave of frequency different from that generated. This phenomenon is known as the *Doppler effect*, and a common acoustical example of it is the change in the pitch of the note of the horn sounded by a passing vehicle.

We shall deduce expressions for the magnitude of the effect. Consider first a source S, initially at rest, emitting a wave of frequency *f*, which propagates with wavelength *λ* and wave velocity *c*, and which is received by an observer O, also initially at rest (Fig. 6.7). We assume that the medium is at rest, and consider motion in only one dimension (i.e. along SO).

When both S and O are at rest, the number of wave crests received by O in unit time is equal to the frequency *f* of the wave. Suppose now that O moves away from S with a constant speed v_O; the number of crests received by O in unit time will be decreased by the number of crests that are contained in a length equal to the distance travelled by O in unit time.

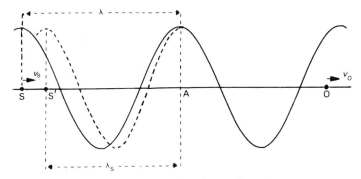

Figure 6.7 Doppler effect in one dimension.

Since this distance is v_O, and the distance between adjacent crests is the wavelength λ, the decrease in the number of crests received by O is v_O/λ. Thus if f_O is the apparent frequency received by O, we have

$$f_O = f - \frac{v_O}{\lambda}.$$

But, since $c = f\lambda$, we have

$$f_O = f - \frac{v_O f}{c}.$$

Thus

$$f_O = f \frac{c - v_O}{c}. \tag{6.46}$$

(We note that this equation predicts that $f_O = 0$ when $v_O = c$, as expected.)

Next we examine the effect of the motion of the source. When the source is at rest, successive wave crests emitted by it are one wavelength λ apart, but when the source is moving towards the observer, the distance between adjacent crests is decreased by the distance the source travels in one cycle. The situation is shown in Fig. 6.7 in which S is moving to the right with constant speed v_S; the distance travelled by S in one cycle is SS′ and the wavelength is decreased from SA $(=\lambda)$ for stationary S to S′A $(=\lambda_S)$. The time taken for one cycle is $1/f$ and the velocity of S is v_S; the distance SS′ is thus v_S/f. Therefore

$$\lambda_S = \lambda - \frac{v_S}{f}.$$

But since the medium in which the wave is being propagated is at rest, the wave velocity is unchanged, so that the frequency of the signal received by a stationary observer is

$$f_S = \frac{c}{\lambda_S}.$$

Replacing λ_S and λ by c/f_S and c/f respectively, we have

$$\frac{c}{f_S} = \frac{c}{f} - \frac{v_S}{f};$$

thus

$$f_S = \frac{cf}{c - v_S}. \tag{6.47}$$

These results can be combined as follows to give the frequency f_{O+S} received by the observer when both source and observer are in motion. The moving source gives rise to a wave whose apparent frequency is given by (6.47). If we take f_S to be the frequency of the wave itself, we need not concern ourselves further with the source. The moving observer will receive a wave whose apparent frequency, f_{O+S}, is obtained by substituting f_S for f in the right-hand side of (6.46) giving

$$\boxed{f_{O+S} = f\,\frac{c - v_O}{c - v_S}.} \tag{6.48}$$

This is a good place to study the Doppler effect on the demonstration disk.

Perhaps the most interesting example of the Doppler effect is in astronomy. When the light from a star is examined spectroscopically, it is found to contain the spectra of common terrestrial elements, but the spectral lines are shifted towards the red end of the spectrum (i.e. the values of λ are all greater than those emitted by atoms in a light source in the laboratory). This 'red shift' can be explained by (6.47), for if the star is travelling away from the Earth, it is evident that $\lambda_S > \lambda$. However, the truth is rather more complex than this since, firstly, (6.47) is only approximately true for light waves (for which a relativistic theory is appropriate if the source velocity is not small compared with that of light) and, secondly, there are other agencies which cause a red shift.

EXAMPLE 6.2 Beats arising from the Doppler effect

A source emitting a sinusoidal sound wave of frequency 500 Hz travels towards a wall at a speed 10 m s^{-1} (Fig. 6.8). What is the beat frequency perceived by an observer travelling away from the wall at a speed 20 m s^{-1}, if the speed of sound is 340 m s^{-1}?

Consider the sound travelling directly from S to O. The frequency perceived by O is given by (6.48), where $v_S = -10$ m s^{-1} and $v_O = 20$ m s^{-1}. Hence $f_{O+S} = 500 \times 320/350$ Hz.

Now the observer will also receive the sound reflected from the wall. How is this to be taken into account? Clearly, the reflection at the wall cannot depend on the motion of the source S; therefore the reflected wavelength must equal the incident wavelength.

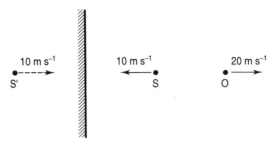

Figure 6.8 Sound source moving towards a wall.

But the incident wavelength is, from the argument given earlier, $\lambda - v_s/f = (c - v_s)/f$. The apparent frequency, if O were stationary, would therefore be $fc/(c - v_s)$. (Note that we are calculating the wavelength to the *left* of S, and therefore that $v_s = +10$ m s^{-1}, not -10 m s^{-1}, for this part of the problem.) Since O is moving, the frequency observed by O is $f(c - v_0)/(c - v_s) = 500 \times 320/330$ Hz.

Although the analysis in the last paragraph is correct, there is a quicker way of arriving at the same result; the observer 'sees' an image S' of the source, behind the wall, moving at a speed 10 m s^{-1} from left to right. The observed frequency from S' is obtained therefore by putting $v_{s'} = 10$ m s^{-1}, $v_0 = 20$ m s^{-1} into (6.48), whence $f_{0+s'} = 500 \times 320/330$ Hz, as previously obtained.

The observer therefore receives two slightly different frequencies, and will perceive an audible 'beat'. The beat frequency is $500 \times 320 \times (1/330 - 1/350) = 27.7$ Hz. ∎

6.10 Dispersion and group velocity

Each physical system treated in this chapter has given a unique wave speed determined solely by the physical constants of the medium concerned. We found that $c = \sqrt{(T/\mu)}$ for a string, and $c = \sqrt{(K/\rho_0)}$ for a fluid, where ρ_0 is the ambient density, so that waves can be propagated with these speeds and with no others.

There are, examples in physics in which the wave velocity turns out to be dependent upon the wavelength, but these are considerably more difficult to treat. One example is that of surface waves on a liquid of depth h, density ρ and surface tension γ for which

$$c^2 = \left[\frac{g}{k} + \frac{\gamma k}{\rho}\right] \tanh(kh),$$

where g is the acceleration due to gravity and k is the circular wavenumber. Another example is that of light waves in a transparent medium, where the relationship between the velocity and wavelength is rather complicated. This property of velocity dependence upon wavelength is called *dispersion*; a medium possessing this property is called a *dispersive medium*.

Of course, a sinusoidal wave will be propagated with an appropriate frequency and wavelength in just the same way as in a non-dispersive medium. More interesting is to examine the collective behaviour of a number of waves of different wavelengths propagated simultaneously through a medium.

Of particular interest is the superposition of sinusoidal waves over a small range of frequencies. The number of component waves may even be infinite (see Chapter 7), but the important requirement is that the frequency range (or bandwidth) is small. The resulting wave is called a *wave group* or *wave packet*, and this moves with a velocity that is quite different from the velocity of the component waves.

To investigate, we analyse the simplest possible example, namely that of just two waves of the same amplitude but of slightly different frequency and circular wavenumber.

Let the two waves be

$$y_1 = a \sin(\omega_1 t - k_1 x)$$

and

$$y_2 = a \sin(\omega_2 t - k_2 x).$$

Then, according to the principle of superposition, the combined disturbance due to these two waves is given by

$$y = y_1 + y_2$$

$$= 2a \sin[\tfrac{1}{2}(\omega_1 + \omega_2)t - \tfrac{1}{2}(k_1 + k_2)x]$$

$$\times \cos[\tfrac{1}{2}(\omega_1 - \omega_2)t - \tfrac{1}{2}(k_1 - k_2)x]. \tag{6.49}$$

The sine term represents a wave whose circular frequency and circular wavenumber are the averages of those of the original wave, and whose wave velocity is $(\omega_1 + \omega_2)/(k_1 + k_2)$. Since we have assumed that ω_1 differs only slightly from ω_2, and k_1 only slightly from k_2, $\tfrac{1}{2}(\omega_1 + \omega_2)$ will differ only slightly from ω_1 or ω_2, and $\tfrac{1}{2}(k_1 + k_2)$ only slightly from k_1 or k_2. Thus the sine term represents a wave whose phase is very similar to those of both the original waves.

The cosine term represents a wave whose circular frequency and circular wavenumber are, respectively, $\tfrac{1}{2}(\omega_1 - \omega_2)$ and $\tfrac{1}{2}(k_1 - k_2)$, and whose velocity is therefore $(\omega_1 - \omega_2)/(k_1 - k_2)$; this term varies more slowly with both time and distance than does the sine term. Figure 6.9 shows how a sketch of the function (6.49) is obtained for fixed t; the ordinates of the sine term (a) and the cosine term (b) are multiplied together, point by point, to produce the heavy curve in (c) which is seen to be contained within an envelope defined by the cosine curve of (b) and its image in the x-axis. It may be seen from the symmetry between the x- and t-terms in (6.49) that a plot of y against t for constant x would give a curve of the same form as that in Fig. 6.9c. The successive building up and dying away of amplitude with time is the phenomenon of beats which has already been referred to in section 2.3.1.

We will examine now how wave groups behave in non-dispersive and dispersive media. In a non-dispersive medium, the wave velocity is constant, so

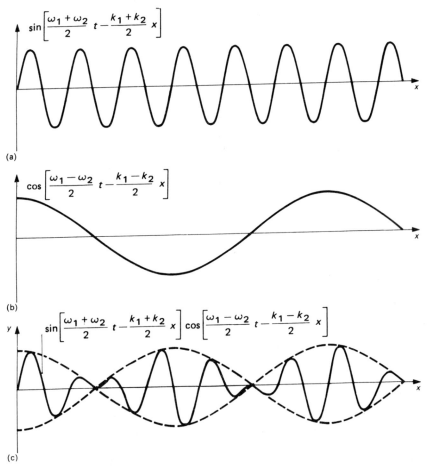

Figure 6.9 Superposition at constant t of two waves of slightly different ω and k: (a) shows the sine term of (6.49), (b) the cosine term and (c) the product of the two terms.

$$c = \frac{\omega_1}{k_1} = \frac{\omega_2}{k_2} = \frac{\omega_1 + \omega_2}{k_1 + k_2} = \frac{\omega_1 - \omega_2}{k_1 - k_2}.$$

This means that the sine part and the cosine part of (6.49) have exactly the same wave velocity, so that as the heavy curve and the envelope of Fig. 6.9c move to the right with increasing time, the position of each relative to the other remains constant. This means that a signal propagated in a non-dispersive medium suffers no change of form.

In the case of a dispersive medium, we have seen that wave velocity varies with wavelength so that

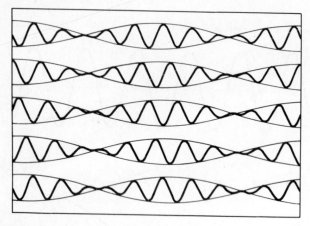

Figure 6.10   Wave group in a dispersive medium. The sketches show the waveform at successive instants of time. It can be seen that the inner curve has advanced a greater distance than the envelope.

$$\frac{\omega_1}{k_1} \neq \frac{\omega_2}{k_2}$$

and therefore

$$\frac{\omega_1 + \omega_2}{k_1 + k_2} \neq \frac{\omega_1 - \omega_2}{k_1 - k_2}.$$

This means that the heavy curve and the envelope in Fig. 6.9c move forward with *different* velocities. The situation is illustrated in Fig. 6.10, where the curves of Fig. 6.9c are shown at successive times. It can be seen in Fig. 6.10 that the inner curve is moving forward at a faster rate than that of the envelope, which means that in this example

$$\frac{\omega_1 - \omega_2}{k_1 - k_2} < \frac{\omega_1 + \omega_2}{k_1 + k_2}. \tag{6.50}$$

It happens that in the great majority of cases in physics, (6.50) holds so that this type of dispersion is referred to as *normal dispersion*; an example is deep-sea waves with wavelengths so large that surface-tension effects can be neglected. On the other hand, if the envelope moves faster than the heavy curve, we have

$$\frac{\omega_1 - \omega_2}{k_1 - k_2} > \frac{\omega_1 + \omega_2}{k_1 + k_2}.$$

This happens less frequently in physics and is referred to as *anomalous dispersion*; examples of it are transverse waves in a solid rod and electromagnetic waves near what is known as an 'absorption edge'.

An extremely important quantity, which we must now examine, is the velocity with which energy is carried forward when two waves are superposed to form a group. When we have a single wave, the energy is carried forward with the velocity at which an amplitude maximum moves forward, which, of course, is the wave velocity. In the case of a wave group, however, we can see from a study of Fig. 6.10 that the velocity with which an amplitude maximum moves forward is that of the *envelope*. It follows that the energy is borne forward with the velocity of the envelope. This velocity is known as the *group velocity* c_g. We saw earlier that the envelope moves with velocity

$$\frac{\omega_1 - \omega_2}{k_1 - k_2},$$

so

$$c_g = \frac{\omega_1 - \omega_2}{k_1 - k_2}. \tag{6.51}$$

We assumed at the outset that ω_1 differed from ω_2, and k_1 from k_2, by small amounts; we may therefore rewrite (6.51) as

$$c_g = \frac{\Delta\omega}{\Delta k}$$

where $\Delta\omega = \omega_1 - \omega_2$ and $\Delta k = k_1 - k_2$. In the limit, as $\Delta k \to 0$ we have

$$\boxed{c_g = \frac{d\omega}{dk}} \tag{6.52}$$

Since $k = 2\pi/\lambda$, this may be written as

$$c_g = \frac{df}{d(1/\lambda)} = -\lambda^2 \frac{df}{d\lambda},$$

i.e.

$$c_g = \frac{1}{2\pi} \frac{d\omega}{d(1/\lambda)} = -\frac{\lambda^2}{2\pi} \frac{d\omega}{d\lambda}. \tag{6.53}$$

Equation (6.52) is one of three useful equivalent expressions for the group velocity; the others can be obtained as follows. Firstly, if we replace ω in (6.52) by kc, where c is the *wave* velocity (the waves which are superposed to form the group have velocities so near to one another that the single value c can be used), then (6.52) becomes

$$c_g = \frac{d(kc)}{dk} = c + k \frac{dc}{dk}. \tag{6.54}$$

Secondly, if we replace k by $2\pi/\lambda$ in the right-hand side of (6.54) we obtain another useful expression,

$$c_g = c + \frac{2\pi}{\lambda}\frac{dc}{d(2\pi/\lambda)},$$

i.e.

$$c_g = c - \lambda\frac{dc}{d\lambda}.$$

The above results have been deduced from the superposition of only two waves, but they are valid for a group comprising a superposition of any number – even an infinite number – of waves.

EXAMPLE 6.3 Vibrations of a one-dimensional crystal

A simple model of a crystal is one in which the atoms are in fixed positions on the sites of a regular lattice. In practice, the atoms will be vibrating about their equilibrium positions because of thermal excitation. Moreover, if one face of the crystal is forced into oscillation by some external means, a wave will travel through the crystal.

A simplified way of analysing such a wave is to consider a one-dimensional crystal, i.e. a linear chain of atoms, each of mass m and separated by a distance a in equilibrium (Fig. 6.11).

We consider the motion of the lth atom, of which the equilibrium x-coordinate is la. The interatomic forces can be modelled by connecting springs, each of stiffness K. Suppose that the atoms suffer arbitrary displacements from their equilibrium positions, that of the lth atom being u_l. The spring to its right has been stretched by an amount $u_{l+1} - u_l$, hence the increase in its tension is $K(u_{l+1} - u_l)$. Likewise, the spring to its left experiences a tension increase of $K(u_l - u_{l-1})$.

The net force on the atom is therefore $K(u_{l+1} - 2u_l + u_{l-1})$, hence its equation of motion by Newton's second law, is

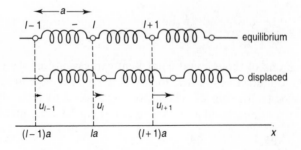

Figure 6.11 Vibrations of a linear chain of atoms.

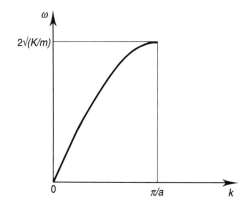

Figure 6.12 Dispersion curve for a one-dimensional chain of atoms.

$$m \frac{\mathrm{d}^2 u_l}{\mathrm{d}t^2} = K(u_{l+1} - 2u_l + u_{l-1}). \tag{6.55}$$

Suppose that a sinusoidal (longitudinal) wave is propagated along the line. It may be represented (section 4.3) by

$$u_l = u_0 \exp \mathrm{i}(\omega t - kla),$$

since the x-coordinate of the lth atom is approximately la, if the amplitude of the vibrations is small. Substituting in (6.55), we have

$$-m\omega^2 = K(\mathrm{e}^{-\mathrm{i}ka} - 2 + \mathrm{e}^{\mathrm{i}ka}) = K(\cos ka - \mathrm{i} \sin ka - 2 + \cos ka + \mathrm{i} \sin ka)$$

$$= 2K(\cos ka - 1) = -4K \sin^2 \tfrac{1}{2}ka.$$

Hence

$$\omega = 2\sqrt{(K/m)} \sin \tfrac{1}{2}ka \tag{6.56}$$

which is illustrated in Fig. 6.12.

This is an example of dispersion, since ω is not proportional to k. In fact, the wave velocity is

$$c_{\mathrm{w}} = \frac{\omega}{k} = 2 \sqrt{\left(\frac{K}{m}\right)} \frac{\sin \tfrac{1}{2}ka}{k} \tag{6.57}$$

and the group velocity is

$$c_{\mathrm{g}} = \frac{\mathrm{d}\omega}{\mathrm{d}k} = a\sqrt{(K/m)} \cos \tfrac{1}{2}ka. \tag{6.58}$$

For small values of ω and k, i.e. near the origin, the graph is approximately straight, and the behaviour is therefore non-dispersive. From (6.57) and (6.58), since $\sin \tfrac{1}{2}ka \approx \tfrac{1}{2}ka$, and $\cos \tfrac{1}{2}ka \approx 1$,

$$c_w \approx c_g \approx a\sqrt{\frac{K}{m}} \tag{6.59}$$

Waves of long wavelength (small k) propagate the chain as if it were a continuous medium, which is what we would expect for $\lambda \gg a$.

It is interesting to note that there is an upper limit to the frequency of propagation. At this upper limit, $k_{max} = \pi/a$, from the graph, so the minimum wavelength is $\lambda_{min} = 2\pi/k_{max} = 2a$. Hence $a = \lambda_{min}/2$; therefore each atom is vibrating in antiphase with its neighbours. Since this is the maximum phase difference that can exist between adjacent atoms, λ cannot be shorter than $2a$ (here we ignore an issue called *aliasing*), and the circular frequency cannot exceed $2\sqrt{(K/m)}$. ∎

6.11 Solitons

A sinusoidal wave will travel in a dispersive medium without change of shape – it remains sinusoidal. However, a more complicated waveform will have components over a range of wavelengths, all of which travel with different velocities. The result is that the wave profile changes as it progresses.

If the medium is non-linear (a term that will be explained presently), then not even a sinusoidal wave is propagated without change of shape.

An interesting effect can arise if the medium is both non-linear and dispersive. A wave of particular profile may be propagated without change of shape. The effects on the profile due to non-linearity and dispersion cancel out. Such a wave is called a *soliton*.

The most familiar example of a soliton occurs with water waves, and was first investigated by J. Scott Russell on the Edinburgh–Glasgow canal in 1834. He noticed that when a boat suddenly stopped, the disturbed water at the bow end formed a well-defined profile which continued along the canal at about 14 km/h, being about 10 m in length and 30–50 cm in height, gradually becoming dissipated over a distance of about a kilometre.

A similar example, well known in Britain, is the Severn Bore. This is a tidal wave in the true sense, caused by large tides in the Severn estuary. (What is popularly known as a 'tidal wave' is more properly termed a *tsunami*, which has nothing to do with tides.) A large wall of water, up to a few metres in height, proceeds along the estuary, offering great interest to spectators and challenge to surfboarders.

The motion of liquid waves was analysed by Korteweg and de Vries. We assume that the liquid is incompressible, has negligible viscosity and surface tension, and that the motion is not rotational.

Suppose that the undisturbed depth of water is h (Fig. 6.13), and that the height of the wave at any point is $\eta(x, t)$ above the undisturbed level. It can be shown that the wave equation is

$$\frac{\partial \eta}{\partial t} + c_0 \frac{\partial \eta}{\partial x} + \left(\frac{3c_0}{2h}\right)\eta\frac{\partial \eta}{\partial x} + \left(\frac{c_0 h^2}{6}\right)\frac{\partial^3 \eta}{\partial x^3} = 0 \tag{6.60}$$

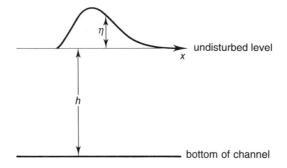

Figure 6.13 A wave on the surface of water in a channel.

where $c_0 = \sqrt{(gh)}$. This is known as the Korteweg de Vries (or KdV) equation. We now examine the various terms, and their effects upon the wave as it progresses.

The first two terms alone would allow any wave profile to be propagated in the positive x-direction with velocity c_0, as was shown in Chapter 4, (4.6). The third term is the non-linear contribution. If η is doubled (i.e. the height of the wave above the undisturbed level is doubled), the other terms are all doubled (they are linear), but the third term is quadrupled.

By writing the first three terms in the form

$$\frac{\partial \eta}{\partial t} + \left(c_0 + \frac{3c_0\eta}{2h} \right) \frac{\partial \eta}{\partial x} = 0$$

(ignoring the fourth term), we see that the effective wave velocity is $c_0 + 3c_0\eta/2h$. This increases with η, indicating that the higher points on the profile move with greater velocity. The effect of this can be seen in Fig. 6.14, which is a sketch of the wave at successive times. As time proceeds, the leading side becomes ever steeper, reverses its slope and leads to a waveform which is clearly unstable.

The fourth term in (6.60) is linear, but dispersive. To see this, consider a sinusoidal wave $\eta = \eta_0 \sin(\omega t - kx)$. Ignoring the non-linear third term, we find by substitution in (6.60) that

$$\omega - c_0 k + \frac{c_0 h^2}{6} k^3 = 0$$

so the wave velocity

$$c = \frac{\omega}{k} = c_0 - \frac{c_0 h^2}{6} k^2.$$

Since c depends on the circular wavenumber k, we have justified our assertion that the fourth term in (6.60) is dispersive. Waves of small wavelength (large k) will travel more slowly than waves of long wavelength.

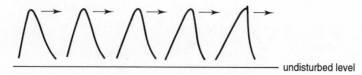

undisturbed level

Figure 6.14 Progress of a water wave in a channel. The non-linear term of (6.60) is included, but the dispersive term is ignored.

We leave it as a problem (6.22) at the end of the chapter to show that for a wave with a Gaussian profile, the higher points on the leading edge move with *smaller* velocity than the lower points. Thus the effects of non-linearity and dispersion on the leading edge oppose each other, and it is easy to appreciate that there is a profile which can be propagated without change of shape – a soliton.

A detailed analysis reveals that the waveform of the soliton is bell-shaped, of the form

$$\eta = \eta_0 \operatorname{sech}^2 \frac{x - ct}{L}, \tag{6.61}$$

where η_0 is the peak value of η, the soliton velocity is

$$c = c_0(1 + \eta_0/2h) \tag{6.62}$$

and

$$L = \left(\frac{4h^3}{3\eta_0}\right)^{1/2} \tag{6.63}$$

The mathematically persistent and adventurous reader is invited to verify that (6.61) satisfies the KdV equation (6.60), (problem 21).

We plot some solitions in Fig. 6.15 for various values of η_0. The larger η_0, the smaller is L, and hence the smaller the width of the soliton. Note too from (6.62) that tall solitons travel faster than short ones.

How do solitons arise? Suppose that we have an initial disturbance which is 'sufficiently localised' and of non-negative volume (e.g. the disturbance may form a hill, but not a depression). Then a full analysis shows that one or more solitary waves emerge, each reaching a stable state with a profile given by (6.61). They emerge in descending order of height, since the tallest are also the fastest, from (6.62).

One very interesting fact concerns a soliton of large height travelling behind one of smaller height (and therefore slower), travelling in the same direction. The former catches up with, and passes through, the latter. Both solitons emerge with no change in shape, which is somewhat surprising in view of the non-linear term in (6.60). The same is true for two solitons travelling in opposite directions and colliding. They pass through each other with no change in shape.

Fascinating experiments to investigate soliton behaviour can be performed with a tank of water a few metres in length. For further details, see A. Bettini, T. A. Minelli and D. Pascoli, 1983, *American Journal of Physics*, **51**, 977–984.

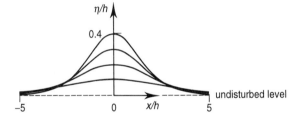

Figure 6.15 Some soliton water waves. $\eta_0/h = 0.1, 0.2, 0.3$ and 0.4.

The accompanying computer program contains a demonstration of a water wave obeying the KdV equation (6.60). The propagation of the wave is shown (a) with the non-linear and dispersive terms ignored, (b) with the non-linear term included and (c) with all terms. A major difficulty in writing the program was in minimising the cumulative errors which arise in solving differential equations numerically. This problem rendered the demonstration of the effect of including the dispersive term rather unconvincing.

This section has concentrated on water-wave solitons for pedagogical reasons. There are many other examples of solitons in physics, and the subject is currently one of intense research activity, particularly in the fields of particle physics, quantum mechanics and fibre optics. Further discussion is beyond the scope of this book.

Problems

1. An elastic string has mass 1 g and natural length 0.1 m. A mass 1 kg is attached to its lower end, and the string is stretched by 0.02 m. Calculate the speed of propagation of transverse waves along the string.

A transverse disturbance travels along the string, and is reflected at both ends. How many times will this disturbance pass a point on the string in one period of a small vertical oscillation of the mass?

(For simplicity, take the tension in the string to be uniform in this problem.)

2. A wire of mass m and length l is fixed at one end, and hangs freely under its own weight. Deduce the time for a transverse wave to travel the length of the wire.

3. A longitudinal wave of amplitude A and circular frequency ω travels with speed c in an elastic rod of density ρ. Assuming that the kinetic and potential energies are equal, prove that the intensity is $\rho A^2 \omega^2 c/2$.

4. Calculate the speed of sound in helium at $-100°C$ ($R = 8.31$ J K^{-1} mol^{-1}, molar mass of He = 4 g).

5. A plane sinusoidal sound wave of displacement amplitude 1.2×10^{-3} mm and frequency 680 Hz is propagated in an ideal gas of density 1.29 kg m^{-3} and pressure 10^5 N m^{-2}. The ratio of principal specific heat capacities is 1.41. Find the pressure amplitude of the wave.

6. It is believed by some people that a sinking ship never reaches the bottom of the ocean, but remains suspended in equilibrium at a certain depth. The 'reason' is that the pressures are enormous, even at modest depths. Using the data that at 25°C, the speed of sound in seawater is 1531 m s^{-1}, and the density of seawater is 1.025×10^3 kg m^{-3}, show that the belief is incorrect as follows: (i) calculate the bulk modulus of seawater; (ii) hence calculate the density at a depth of 8 km; (iii) show that an iron ship will certainly sink to the bottom of the ocean.

What assumptions have been made in the analysis?

7. Calculate the speed of longitudinal waves in steel (Young modulus $= 2.0 \times 10^{11}$ N m^{-2}, density $= 7.8 \times 10^3$ kg m^{-3}).

8. (harder) Justify the assumption that the pressure changes in a gas carrying a sound wave are adiabatic as follows.

The regions of compression (at temperature T_1) and rarefaction (at T_2) can be thought of as occupying a length of, say, $\sim \lambda/4$, and as separated by $\sim \lambda/2$. Defining θ as $T_1 - T_2$, and K as the thermal conductivity, show that $\mathrm{d}\theta/\theta = -\mathrm{d}t/\tau_r$, where $\tau_r \sim \lambda^2 \rho c_p/K$ (c_p refers to unit mass). By integration, show that τ_r is the 'relaxation time' which determines how quickly the gas attains a uniform temperature.

Hence show that for a sound wave of frequency f and period τ, the ratio $\tau/\tau_r \sim Kf/c^2 \rho c_p$. Show that this ratio for typical frequency of 1 kHz in air is small, given that $K = 0.023$ W m^{-1} K^{-1}, $c = 340$ m s^{-1}, $\rho = 1.2$ kg m^{-3}, $C_{p,M} = 7R/2$ (for 1 mole), $R = 8.31$ J K^{-1} mol^{-1} and the molar mass of air = 29 g mol^{-1}.

(It is interesting to note that the adiabatic behaviour is better obeyed at the *lower* frequencies.)

9. A length of coaxial cable is used to connect a television aerial to the receiving set. The inductance and capacitance per unit length of the cable are respectively 5×10^{-7} H m^{-1} and 8×10^{-11} F m^{-1}. If the frequency of the broadcast signal is 600 MHz, calculate (i) the wavelength in free space and (ii) the wavelength in the cable.

10. Show that if a sinusoidal wave is propagated along a transmission line in one direction, the voltage V and the current i are in phase or antiphase at all points, and that $|V/i| = \sqrt{(L/C)}$. This quantity is called the characteristic impedance of the line Z_0.

Calculate Z_0 if $L = 7 \times 10^{-7}$ H m^{-1} and $C = 6 \times 10^{-11}$ F m^{-1}.

11. A car is travelling along a road on which there is a speed limit of 70 m.p.h. ($=31.3$ m s^{-1}). A stationary policeman notice that the pitch of the horn falls by an interval of a major third (i.e. a frequency ratio of 5/4) when the car passes. To what conclusion might he come? (Speed of sound in air $= 340$ m s^{-1}.)

12. Show that it is possible for zero frequency to be perceived if the observer is in motion, but not if the source is in motion with the observer stationary.

13. How would equation (6.48) be modified if there were a wind moving from observer to source with velocity v_w?

14. Show that if v_s and v_o are both small compared with c, the fractional change in frequency due to the Doppler effect depends on the *relative* velocity, and is in fact $(v_s - v_o)/c$.

15. Using the result of problem 14, and assuming (correctly!) that it also applies to the Doppler effect for light, estimate the Doppler spread in wavelength from hydrogen atoms which, if stationary, would emit light with wavelength 650 nm, but are in fact moving randomly with speeds $\sim 3 \times 10^3$ m s^{-1} (speed of light $= 3 \times 10^8$ m s^{-1}).

16. (longer) Consider the meaning of the 'apparent wavelength of a stationary source to a moving observer'. This is less well-defined than the frequency, which is unambiguous. Consider two ways in which the wavelength may be measured: (i) by taking an instantaneous 'photograph' of the medium with its troughs and crests and (ii) by performing an acoustic Young's slits experiment (Chapter 10).

No solution is given, but the interested reader is referred to *Physics Education* **16**, 366–8, 1981.

17. Deep-water waves are defined as being such that λ is much smaller than the depth of water, but large enough for surface tension effects to be negligible. Show that for such waves, $c^2 = g\lambda/2\pi$. [N.B. tanh $\theta = (e^\theta - e^{-\theta})/(e^\theta + e^{-\theta})$.]

Calculate the wave velocity if $\lambda = 10$ m.

Show that the group velocity of deep-water waves is half the wave velocity.

18. Show that for shallow water waves (i.e. λ is much greater than the depth h, and surface tension effects are negligible), $c^2 = gh$. Are such waves dispersive?

What is the group velocity?

19. Surface ripples on water are such that λ is much smaller than the depth, and also so small that surface tension effects dominate (i.e. $2\pi\gamma/\rho\lambda \gg g\lambda/2\pi$). What range of wavelengths would satisfy this condition for water? ($\gamma = 0.073$ N m^{-1}.)

Show that for such waves, $c^2 = 2\pi\gamma/\rho\lambda$. Calculate the wave velocity if $\lambda = 1$ mm.

Show that the group velocity is 3/2 times the wave velocity. (Note that this is a case of anomalous dispersion.)

20 Deduce an expression for the wave velocity c as a function of λ for a hypothetical medium in which the group velocity is a constant V at all wavelengths. Show that the only physically acceptable solution is one in which the medium is non-dispersive.

21. Verify that $\eta = \eta_0 \operatorname{sech}^2 (x - ct)/L$ satisfies the Korteweg de Vries equation (6.60), where c and L are as given in the text. [Note: $\operatorname{sech} \theta = 1/\cosh \theta$, $d(\cosh \theta)/d\theta = \sinh \theta$, $d(\sinh \theta)/d\theta = \cosh \theta$ and $\cosh^2 \theta - \sinh^2 \theta = 1$.]

22. Consider a water wave with Gaussian profile $\eta = \eta_0 \exp(-\alpha x^2)$ where η_0 and α are positive constants. Argue that the higher points on the leading edge of the wave move with smaller velocity than the lower points when the non-linear term is ignored and the dispersive term included. [Hint: deduce the ratio of the terms in $\partial^3 \eta/\partial x^3$ and $\partial \eta/\partial x$, and show that this increases with x (for positive x).]

7 Fourier series and transforms

7.1 Introduction and mathematical discussion

The theorem of J. B. J. Fourier, in various guises, is an important proposition in physics. It is probably true to say that there is no branch of the subject which this theorem has not illuminated in some significant way or other. Although it finds its main use in studies involving wave motion, Fourier first introduced it (in 1822) in connection with the problem of heat conduction, in which subject it still occupies a prominent position. These two applications themselves illustrate the diversity of situations to which it has been applied.

Fourier's theorem is essentially a trigonometrical relationship which can be applied to a large class of mathematical functions, of which periodic functions provide the simplest examples. We will deal with periodic functions first, and later on in this chapter go on to non-periodic functions.

Mathematically, a periodic function of time $g(t)$, of period T, is a function which has the property that

$$g(t + T) = g(t),$$

for all values of t. By a simple iteration of this expression one can see that

$$g(t + NT) = g(t),$$

for all t, where N is any integer, positive, negative or zero. It necessarily follows that the function extends infinitely along the positive and negative t axes.

Figure 7.1 illustrates some periodic functions, each of the same period T; in particular, Fig. 7.1a illustrates the function

$$y(t) = A \cos\left(\frac{2\pi t}{T} + \phi\right)$$

plotted against time. The profile of this is sinusoidal, the amplitude is A and the phase constant ϕ determines how far the first maximum is away from the origin. If we substitute f_0 for $1/T$, the equation becomes

$$y(t) = A \cos(2\pi f_0 t + \phi). \tag{7.1}$$

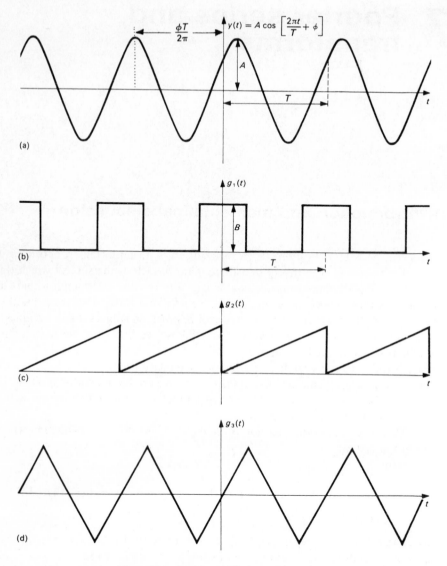

Figure 7.1 Some periodic functions of time.

The quantity f_0, being the reciprocal of T which is the time duration of one period, is thus the number of complete periods per unit time. It is known as the *fundamental frequency* of the periodic function.

Now Fourier's theorem states that *any* periodic function $g(t)$ can be expressed as the sum (to an infinite number of terms if necessary) of functions of the type appearing in (7.1), where the frequencies appropriate to each term in the sum are integral

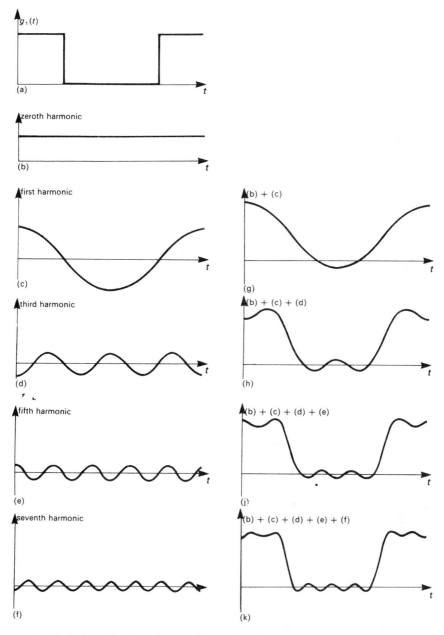

Figure 7.2 Illustration of Fourier's theorem. One period of the function illustrated in Fig. 7.1b appears in (a). The curves (b)–(f) on the left-hand side show the first five frequency components into which (a) is analysed. On the right-hand side, (g)–(k), are shown successive stages in the addition of the components in the synthesis of the original function. The last curve (k) is seen to be a reasonable approximation to (a) after the addition of only five terms.

multiples, nf_0, of the fundamental f_0 and the amplitudes A and phases ϕ are, in general, different for different values of n. The theorem implies that any periodic function of time can be *analysed* into *frequency components*; alternatively the original function can be *synthesised* by summing these frequency components. Figure 7.2 illustrates these processes of analysis and synthesis for the periodic function shown in Fig. 7.1b.

Expressed in mathematical terms, Fourier's theorem asserts that, for a periodic function of fundamental frequency f_0,

$$g(t) = \sum_{n=0}^{\infty} A_n \cos(2\pi nf_0 t + \phi_n), \tag{7.2}$$

where the suffixes on the A's and the ϕ's denote that these latter belong to a particular n.

If we wish to Fourier-analyse a given periodic function, our problem essentially is to determine the values of the amplitudes A_n, and phases ϕ_n, for each of its frequency components. We begin by expressing a typical term of the sum, namely $A_n \cos(2\pi nf_0 t + \phi_n)$, as

$$A_n \cos(2\pi nf_0 t) \cos \phi_n - A_n \sin(2\pi nf_0 t) \sin \phi_n$$

$$= (A_n \cos \phi_n) \cos 2\pi nf_0 t - (A_n \sin \phi_n) \sin 2\pi nf_0 t,$$

where those terms that are independent of time have been collected together in brackets. For convenience we can replace these terms respectively by C_n and $-S_n$, so the expression becomes

$$C_n \cos 2\pi nf_0 t + S_n \sin 2\pi nf_0 t.$$

Thus, from (7.2),

$$g(t) = \sum_{n=0}^{\infty} C_n \cos 2\pi nf_0 t + \sum_{n=0}^{\infty} S_n \sin 2\pi nf_0 t. \tag{7.3}$$

Our problem of finding the A_n and ϕ_n appropriate to a particular periodic function $g(t)$ has now become that of finding the C_n and S_n.

Now, concentrating our attention on a particular value of n, say m, we multiply (7.3) throughout by $\cos 2\pi mf_0 t$ and integrate with respect to t over a whole period, obtaining

$$\int_{-1/2f_0}^{+1/2f_0} \cos(2\pi mf_0 t)g(t)\, dt = \int_{-1/2f_0}^{+1/2f_0} \cos(2\pi mf_0 t)\left(\sum_{n=0}^{\infty} C_n \cos 2\pi nf_0 t \right) dt$$

$$+ \int_{-1/2f_0}^{+1/2f_0} \cos(2\pi mf_0 t)\left(\sum_{n=0}^{\infty} S_n \sin 2\pi nf_0 t\, dt \right) \tag{7.4}$$

where the range of the integral (from $-1/2f_0$ to $+1/2f_0$) has been expressed in terms of f_0. Equation (7.4) appears very long and cumbersome, but we shall now see that

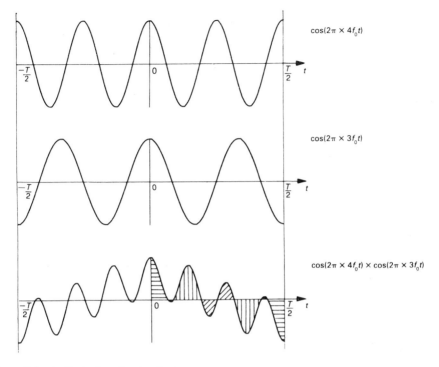

Figure 7.3 An illustration showing that

$$\int_{-1/2f_0}^{+1/2f_0} \cos(2\pi \times 3f_0 t)\, \cos(2\pi \times 4f_0 t)\, dt = 0.$$

remarkable simplifications can be effected on the right-hand side. A typical term in the first integral on the right-hand side is

$$C_n \int_{-1/2f_0}^{+1/2f_0} \cos(2\pi m f_0 t)\, \cos(2\pi n f_0 t)\, dt \tag{7.5}$$

and similarly a typical term in the second integral is

$$S_n \int_{-1/2f_0}^{+1/2f_0} \cos(2\pi m f_0 t)\, \sin(2\pi n f_0 t)\, dt, \tag{7.6}$$

the quantities C_n and S_n being brought outside the integral sign because they are constants. Let us take (7.5) first, and suppose $m \neq n$. A little thought will reveal that the value of the integral is zero. Figure 7.3 illustrates this for the case $m = 4$ and $n = 3$, where the first two graphs are those of the two individual cosine terms and the third graph is that of their product. Half the area of the third graph has been shaded with four different shadings; it will be noticed that each type of shading occurs twice, once for a positive area (above the time axis) and once for an equal negative area (below the axis). All the positive areas therefore have equal negative counterparts, producing a zero overall value. The integral (7.5) can, of course, be carried out analytically; the reader might care to try this.

Continuing with our consideration of the expression (7.5), and taking the case where $n = m$, we see that the expression now becomes

$$C_m \int_{-1/2f_0}^{+1/2f_0} \cos^2 2\pi m f_0 t \, dt = \frac{C_m}{2} \int_{-1/2f_0}^{+1/2f_0} (1 + \cos 4\pi m f_0 t) \, dt$$

$$= \frac{C_m}{2} \left[t + \frac{\sin 4\pi m f_0 t}{4\pi m f_0} \right]_{-1/2f_0}^{+1/2f_0}$$

$$= \frac{1}{2f_0} C_m,$$

provided m is not zero. If m is zero, then the expression becomes

$$C_0 \int_{-1/2f_0}^{+1/2f_0} dt$$

which is $(1/f_0)C_0$.

Coming now to expression (7.6) we have to evaluate

$$\int_{-1/2f_0}^{+1/2f_0} \cos(2\pi m f_0 t) \sin(2\pi n f_0 t) \, dt,$$

which can easily be shown to be equal to zero regardless of whether $m = n$ or not. (It is suggested that the reader try this.)

So (7.4), which initially looked rather formidable, has been reduced to

$$\int_{-1/2f_0}^{+1/2f_0} g(t) \cos(2\pi m f_0 t) \, dt = \frac{1}{2f_0} C_m$$

when $m \neq 0$, and to

$$\int_{-1/2f_0}^{+1/2f_0} g(t) \, dt = \frac{1}{f_0} C_0$$

when $m = 0$.

This gives us values for the C_m as follows:

$$C_0 = f_0 \int_{-1/2f_0}^{+1/2f_0} g(t) \, dt,$$

$$C_m = 2f_0 \int_{-1/2f_0}^{+1/2f_0} g(t) \cos 2\pi m f_0 t \, dt, \; m \neq 0. \qquad (7.7)$$

If we had started from (7.3) by multiplying by $\sin 2\pi mf_0 t$ instead of $\cos 2\pi mf_0 t$, we would, by a precisely similar argument, have arrived at the conclusion that

$$S_m = 2f_0 \int_{-1/2f_0}^{+1/2f_0} g(t) \sin 2\pi mf_0 t \, dt. \tag{7.8}$$

(So, unlike C_0, S_0 is always zero, since the integral clearly vanishes for $m = 0$.) Equations (7.7) and (7.8) tell us how to find the coefficients C_n and S_n corresponding to a given periodic function $g(t)$, and hence enable us to perform the frequency analysis of any such function.

While the limits in all the integrals above have been $-1/2f_0$ and $+1/2f_0$, there is, actually no need for this to be so. It is merely necessary that they be one period apart; some functions may be more easily Fourier-analysed with limits which are not symmetric.

EXAMPLE 7.1

Let us take as a concrete example a function of period $T(=1/f_0)$ which has a value of B in the range $-\alpha T/2 < t < +\alpha T/2$ and of zero elsewhere in the range $-T/2$ to $+T/2$. Here, α is the proportion of the period for which the function has the non-zero value of B. The function is illustrated in Fig. 7.1b for $\alpha = \frac{1}{2}$. Equation (7.7) for $m = 0$ yields immediately $C_0 = \alpha B$. For $m \neq 0$, (7.7) becomes

$$C_m = 2f_0 \int_{-\alpha T/2}^{\alpha T/2} B \cos(2\pi mf_0 t) \, dt$$

$$= 2f_0 B \left[\frac{\sin 2\pi mf_0 t}{2\pi mf_0} \right]_{-\alpha T/2}^{\alpha T/2}$$

$$= \frac{2B}{\pi m} \sin(\pi m\alpha).$$

Taking, as an example, $\alpha = 3/4$ and substituting $m = 1, 2, 3, \ldots$ into this last result, we obtain

$$C_1 = \frac{2B}{\pi\sqrt{2}}, \quad C_2 = -\frac{B}{\pi}, \quad C_3 = \frac{2B}{3\pi\sqrt{2}}, \quad C_4 = 0,$$

$$C_5 = -\frac{2B}{5\pi\sqrt{2}}, \quad C_6 = \frac{2B}{6\pi}, \quad C_7 = -\frac{2B}{7\pi\sqrt{2}} \quad \text{and} \quad C_8 = 0.$$

Having now found the C_m, let us turn our attention to finding the S_m, from (7.8). With the same limits of integration as before (7.8) gives

$$S_m = 2f_0 \int_{-\alpha T/2}^{+\alpha T/2} B \sin 2\pi mf_0 t \, dt$$

$$= -2f_0 B \left[\frac{\cos 2\pi mf_0 t}{2\pi mf_0} \right]_{-\alpha T/2}^{+\alpha T/2}$$

$$= 0 \quad \text{for all } m. \qquad \blacksquare$$

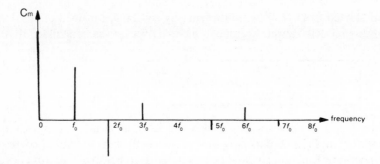

Figure 7.4 Frequency spectrum of the function of period *T*, of value *B* in the range $-\alpha T/2 < t < +\alpha T/2$ and zero elsewhere. (Here, $\alpha = 3/4$.)

In the example just considered, the S_m were all zero. In fact this is true for *any* periodic function which is also *even* – that is to say, any function $g(t)$ where

$$g(-t) = g(t).$$

The function we have taken is clearly even. In a similar way, any odd function, i.e. any function for which

$$g(t) = -g(-t),$$

produces zero C_m. Figure 7.1(d) illustrates such a function.

The C_m for the function which is our example (with $\alpha = 3/4$) can now be displayed in graphical form by vertical lines plotted along an abscissa representing frequency, whose heights are proportional to the C_m. The result is shown in Fig. 7.4. The reader might like to use these values of C_m to synthesise the wave with the disk.

The frequency f_0 (the fundamental) is alternatively known as the *first harmonic*. The frequency $2f_0$ is termed the *second harmonic*; $3f_0$ is the *third harmonic* and so on. In the particular case of $\alpha = \frac{1}{2}$, it is interesting to note that even harmonics have zero amplitude; they are said to be 'missing'.

Before we leave the subject of the mathematical aspects of the Fourier analysis of periodic functions, there is one important step to be taken. It turns out that Fourier's theorem can be recast, with considerable advantage, into a form involving exponentials with imaginary exponents. We will now do this, and see the advantage it brings.

The representation (7.2) of a periodic function with which we started consisted of the sum of cosine functions of different amplitudes, phases and frequencies. Such a function (illustrated in Fig. 7.1a) has the familiar regular undulating shape. The function can, however, by virtue of the identity

$$\cos \theta = \tfrac{1}{2}[\exp(\mathrm{i}\theta) + \exp(-\mathrm{i}\theta)],$$

be itself further analysed into the sum of two components each of which turns out to be mathematically much more tractable and manipulable than the cosine itself. Specifically, the identity reveals that the nth term in (7.2), namely $A_n \cos(2\pi n f_0 t + \phi_n)$ can be put into the form

$$\left[\frac{A_n}{2} \exp(i\phi_n)\right] \exp(2\pi i n f_0 t) + \left[\frac{A_n}{2} \exp(-i\phi_n)\right] \exp[2\pi i(-nf_0)t]. \tag{7.9}$$

The number $(A_n/2) \exp(i\phi_n)$ is a complex number, say G_n, whose amplitude is $A_n/2$ and whose phase is ϕ_n. The number $(A_n/2) \exp(-i\phi_n)$ is the complex conjugate, G_n^*, *of* G_n since its phase is equal and opposite to that of G_n. So (7.9) can be written as

$$G_n \exp(2\pi i n f_0 t) + G_n^* \exp[2\pi i(-nf_0)t]. \tag{7.10}$$

Now, $\exp(2\pi i n f_0 t)$ as a function of t is complex. Its real part is $\cos(2\pi n f_0 t)$ and its imaginary part is $\sin(2\pi n f_0 t)$ so that its magnitude, namely $[\cos^2(2\pi n f_0 t) + \sin^2(2\pi n f_0 t)]^{1/2}$ is unity for all t. Drawn on an Argand diagram against t, it has the appearance of a helix whose axis is the t-axis. The mutually perpendicular real and imaginary axes of the Argand diagram are both at right angles to that of the helix. Alternatively, one may picture $\exp(2\pi i n f_0 t)$ as being on the curved edge of a corkscrew whose axis is the t-axis. The pitch of the helix – the time taken for one turn around the corkscrew – is $1/nf_0$. To see this, note that $\exp(i\theta)$ makes one turn around the θ-axis as θ increases by 2π. So $\exp(2\pi i n f_0 t)$ makes one turn in a time t such that $2\pi n f_0 t = 2\pi$, or $t = 1/nf_0$ as stated. The higher that either n or f_0 is, the lower is the pitch of the helix, i.e. the faster is the helical rate of rotation.

So the function $\exp(2\pi i n f_0 t)$ is of helical shape (on an Argand diagram) with frequency nf_0. The other term, $\exp[2\pi i(-nf_0)t]$ in (7.10) is of identical helical shape but with a rotation in the opposite sense. In other words, it can be pictured as an 'anticlockwise' corkscrew. Its frequency is thus $-nf_0$, which means that we have the concept of a *negative frequency*. Negative frequencies, though impossible to imagine in the context of cosine functions, are thus quite natural if we choose to analyse a periodic function into the helical functions under discussion. In particular, we note that $\cos(2\pi n f_0 t)$, analysed into helical functions, has two components, one at frequency nf_0 and the other at frequency $-nf_0$. Finally, we note that if G_n in (7.10) is the complex amplitude of the component at frequency nf_0, then G_n^* is that of the component at $-nf_0$; it is convenient, therefore, to replace the symbol G_n^* by G_{-n}.

We will now proceed to see the advantages of expressing the periodic function $g(t)$ as

$$g(t) = \sum_{n=-\infty}^{\infty} G_n \exp(2\pi i n f_0 t) \tag{7.11}$$

(i.e. with frequency components nf_0 from $-\infty$ to $+\infty$) rather than the original

$$g(t) = \sum_{n=0}^{\infty} A_n \cos(2\pi n f_0 t + \phi_n), \tag{7.2}$$

noting that the representations are interchangeable with $|G_n| = A_n/2$ and $\arg G_n = \phi_n$.

To obtain the complex coefficients G_n we proceed much as we did previously to obtain S_n and C_n. We multiply (7.11) by

$$\exp(-2\pi i m f_0 t)\, dt,$$

where m is an integer (this time positive, negative or zero), and integrate over a whole period of the function $g(t)$ to obtain

$$\int_{-1/2f_0}^{+1/2f_0} g(t) \exp(-2\pi i m f_0 t)\, dt$$

$$= \int_{-1/2f_0}^{+1/2f_0} \exp(-2\pi i m f_0 t) \sum_{n=-\infty}^{\infty} G_n \exp 2\pi i m f_0 t\, dt. \tag{7.12}$$

A typical term on the right-hand side of (7.12) is

$$G_n \int_{-1/2f_0}^{+1/2f_0} \exp 2\pi i (n-m) f_0\, dt = G_n \left[\frac{\exp 2\pi i (n-m) f_0 t}{2\pi i (n-m) f_0} \right]_{-1/2f_0}^{+1/2f_0}$$

$$= \frac{G_n}{2\pi i (n-m) f_0} \{\exp i\pi (n-m) - \exp[-i\pi (n-m)]\}$$

$$= \frac{G_n}{\pi (n-m) f_0} \frac{\exp i\pi (n-m) - \exp[-i\pi (n-m)]}{2i}$$

and, since $[\exp i\theta - \exp(-i\theta)]/2i = \sin\theta$, this becomes

$$\frac{G_n}{\pi (n-m) f_0} \sin \pi (n-m) = 0 \quad \text{if } n - m \neq 0.$$

So we see, as before, that if $n - m$ is not equal to zero a typical term on the right-hand side of (7.12) is zero. For the case $n - m = 0$, we note from the equation following (7.12) that

$$G_n \int_{-1/2f_0}^{+1/2f_0} dt = \frac{G_n}{f_0}.$$

Thus (7.12) becomes

$$\int_{-1/2f_0}^{+1/2f_0} g(t) \exp(-2\pi i m f_0 t)\, dt = \frac{G_m}{f_0},$$

so the mth coefficient is

$$\boxed{G_m = f_0 \int_{-1/2f_0}^{+1/2f_0} g(t) \exp(-2\pi i m f_0 t)\, dt.} \tag{7.13}$$

Equation (7.13) enables us to obtain all the Fourier coefficients for a given $g(t)$ by just one integration instead of the previous two. The reader may find it instructive to use (7.11) and (7.13) to Fourier-analyse the square wave treated previously. Equation (7.13) is the equation for the complex coefficients G_m corresponding to the two equations (7.7) and (7.8) for the trigonometrical version. We note that (7.13) is valid for any integer m, zero or otherwise, whereas the zeroth term in the trigonometrical version needs to be treated separately from those for which $m \neq 0$.

We finally make the following remark concerning the complex representation (7.11) of a real function $g(t)$. The right-hand side of (7.11) has, of course, to be real in order for it to be equal to the real quantity $g(t)$ on the left-hand side. But how can this be so when each term in the series of (7.11) is complex? The answer lies in the fact that the sum of the nth and $-n$th terms is real for all non-zero n and the term for $n = 0$ is also real.

7.2 The physical significance of Fourier's theorem

Now that we have seen how to analyse a periodic function of time into its frequency components, we must enquire rather more deeply into the significance of the operation. Fourier theory has many applications to mechanical, electrical and other sytems which have the properties of *linearity* and *time-invariance*. The response of the system (which may be displacement, current or some other quantity) to a stimulus (which may be a force or an e.m.f. varying with time) is governed, in general, by a differential equation connecting the response to the stimulus. If this equation is linear, the system is described as *linear* with respect to the quantities being considered; if the coefficients in the equation are constants, independent of time, the system is said to be *time-invariant*. Many systems studied in physics possess these two quantities to very good approximations. As an example, a mechanical system whose displacement can be characterised by a variable $y(t)$ impressed with a sinusoidally varying force $F \sin(2\pi f t + \varepsilon)$ might be described by the equation

$$a \frac{d^2 y}{dt^2} + b \frac{dy}{dt} + cy = F \sin(2\pi f t + \varepsilon),$$

where a, b and c are constants. Such a system would be linear, because of the linearity of the dependent variable y, and time-invariant because of the constancy of the coefficients a, b and c.

If the stimulus $F \sin(2\pi f t + \varepsilon)$ is assumed to have been applied for all time (i.e. $-\infty \leqslant t \leqslant +\infty$) then any transients of the kind described in the Appendix will not manifest themselves. In this case, the solution to the equation is of the form

$$y = A \sin(2\pi f t + \phi),$$

where A and ϕ are constants determined entirely by a, b, c, F, f and ε. This is a most important result. It states that the response y to a force F varying sinusoidally with time *itself* varies sinusoidally with time at the same frequency. Suppose the input force were of the form

$$F_1 \sin(2\pi f_1 t + \varepsilon_1) + F_2 \sin(2\pi f_2 t + \varepsilon_2);$$

then, since the equation is linear, we can use the principle of superposition to show that y is of the form

$$y = A_1 \sin(2\pi f_1 t + \phi_1) + A_2 \sin(2\pi f_2 t + \phi_2)$$

and so on, for any number of frequency components in the stimulus.

Thus an input stimulus of a given frequency spectrum can only produce a response with a spectrum containing the *same frequencies*. No new frequencies can be generated. Each frequency component may be modified in both phase and amplitude, but the frequencies in the output are still those of the input. If we know the Fourier coefficients of the components in the stimulus and also know how the system modifies these, then we can determine the coefficients of the spectral components in the response, and, using (7.11), synthesise the waveform of the response. So, Fourier analysis of a periodic waveform turns out to be a very significant analysis in this case, since a sinusoidal variation has an *invariance*, with respect to a linear system, which no other wave shape possesses.

There are, of course, many other ways of analysing a function. For many purposes a useful way is by expressing the function $g(t)$ as a power series

$$g(t) = a_0 + a_1 t + a_2 t^2 + \dots,$$

or, in a more compact notation,

$$g(t) = \sum_{m=0}^{\infty} a_m t^m.$$

If an input stimulus were analysed in this way, the principle of superposition would still hold in that the response of the system to each individual component of the stimulus in the power series could be determined, and the effects of each term added to synthesise the final response. But the operation would be much more complex; one term in the power series, say $a_3 t^3$, would generate not only the cubic term in the response, but a whole power series, as would all the other components. On the other hand, if the original $g(t)$ had been *Fourier*-analysed, each component frequency in the stimulus would have produced that, and only that, frequency component in the response.

These processes of *frequency* analysis and synthesis are important in, for example, audio systems. When we buy our high-fidelity reproduction equipment, we demand of the manufacturer that the amplifier and loudspeaker be able to transmit the entire audio range of frequencies to our ears. Since we can hear sounds of frequencies up to about 15 000 Hz, and our tapes and CDs contain frequencies up to that value, we require that the system does not modify the amplitude of the Fourier components in that range.

Fourier analysis is, then, a most useful concept in wave theory, and before passing on to more applications of it we must see how the theory can be generalised still further to include non-periodic signals.

7.3 The Fourier transform

Many of the uses to which waves are put are in the field of communications; for example, light waves communicate information to our eyes, sound waves to our ears and radio waves allow information to be broadcast over large distances. It is immediately evident that the information being transmitted is not, in general, a periodically repeating function; this section will deal with the frequency analysis of those signals which are not of the simple periodic type so far described in this chapter.

There are many approaches to this problem; a rigorous approach is mathematically very complex and far outside the scope of this book. Instead, we shall give an exposition which will indicate the plausibility of the theory, but which will probably not appeal to the mathematical purist.

We start with (7.11), which is the complex form of Fourier's theorem for periodic functions,

$$g(t) = \sum_{n=-\infty}^{\infty} G_n \exp(2\pi i n f_0 t),$$ [7.11]

together with the equation (7.13), for determining the coefficients G_n of the series,

$$G_n = f_0 \int_{-1/2f_0}^{+1/2f_0} g(t) \exp(-2\pi i n f_0 t)\, dt,$$ [7.13]

where n has been substituted for m. Let us consider (7.11). A non-periodic function can be thought of as a periodic function of infinitely long period (and therefore of infinitesimal fundamental frequency f_0). As the period gets longer and longer, the component frequencies become closer and closer together on the frequency scale. For example, if the period were as long as 1 s, the fundamental f_0 would be 1 Hz and each component would be 1 Hz apart from its neighbours. If the period is now extended to 100 s, the fundamental would be 1/100 Hz and the components would be only 1/100 Hz apart in frequency. And so, as we let the period tend to infinity, the components become infinitely close together and in the limit merge to form a continuum. Instead of the discrete frequencies nf_0, we have a continuous variable f; likewise, the G_n in (7.11) are replaced by a quantity which describes the total amplitude within an elementary frequency band of width df. Let us call this quantity $G(f)\, df$, where $G(f)$ is the total amplitude per unit frequency range at the frequency f. The summation sign now becomes an integral and (7.11) becomes

$$g(t) = \int_{-\infty}^{\infty} G(f) \exp(2\pi i f t)\, df.$$ (7.14)

So the spectrum of a non-periodic function $g(t)$ is a *continuous* spectrum described by $G(f)$ and, if we know the spectrum, we can reconstruct the function $g(t)$ from it by the integral in (7.14).

But suppose that we know the original function $g(t)$ and wish to find its spectrum. To do this we must turn to (7.13) for periodic functions and modify it in a way similar to that employed for (7.11). If we consider the right-hand side of (7.13) as the period of $g(t)$ tends to infinity, then nf_0 is replaced (as before) by the continuous variable f, f_0 becomes df and the limits of the integral, $+1/2f_0$ and $-1/2f_0$, tend to $+\infty$ and $-\infty$ respectively. Since G_n on the left-hand side of (7.13) is replaced by $G(f)\,df$, this equation now becomes

$$G(f) = \int_{-\infty}^{+\infty} g(t)\exp(-2\pi ift)\,dt. \qquad (7.15)$$

This is the equation we employ to find the spectrum of a non-periodic function $g(t)$.

The two equations (7.14) and (7.15) represent a very general description of Fourier's theorem for functions of time. They also illustrate the essential symmetry which underlies the theorem, in that (7.15) is identical to (7.14) except that the variables t and f have been interchanged, the two functions G and g have likewise been interchanged, and the exponent is negative in one case and positive in the other. Any function $g(t)$ (with certain reservations which need not concern us here) when integrated to give its spectrum (7.15), which is then integrated again (7.14), will yield precisely the original function $g(t)$. This shows the very important point that if the spectrum $G(f)$ of a signal $g(t)$ is known completely, then the original signal $g(t)$ can be determined uniquely; if this were not so, then the process of integrating twice as described above would not be capable of generating the original function. All the *information* in the original signal is therefore preserved in its spectrum.

The function $G(f)$ is known as the *Fourier transform* (FT; or *Fourier integral*) of the function $g(t)$. In general, as (7.15) shows, $G(f)$ is a complex function, even if the signal $g(t)$ is real. In fact, for real $g(t)$, $G(f)$ is *Hermitian*, i.e. $G(f) = G^*(-f)$ where the asterisk denotes the complex conjugate. The proof of this result forms one of the problems at the end of this chapter. We concentrate here on simpler, but similar, examples. First, we show that $G(-f)$ is the FT of $g(-t)$, i.e. that if one turns the signal 'back to front', the same thing happens to its FT. To show this, substitute $-f$ for f in (7.15) and obtain

$$G(-f) = \int_{-\infty}^{\infty} g(t)\exp[-2\pi i(-f)t]\,dt.$$

Now substitute $-t$ for t and obtain

$$G(-f) = \int_{+\infty}^{-\infty} g(-t)\exp[-2\pi i(-f)(-t)]\,d(-t).$$

(Note the inversion of the integral limits; when $t = -\infty$, then $-t = +\infty$.)

$$G(-f) = \int_{-\infty}^{+\infty} g(-t)\exp(-2\pi ift)\,dt.$$

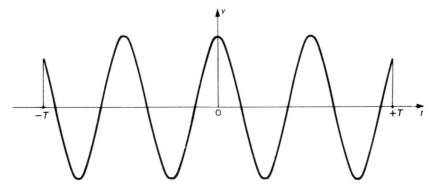

Figure 7.5 Cosine function extending over a finite time.

This last integral tells us that $G(-f)$ is the FT of $g(-t)$ as asserted: A corollary of this result and the Hermitian property is that if $g(t)$ is real and even (i.e. $g(t) = g(-t)$) then so is $G(f)$. We show this as follows. For any real $g(t)$, we have $G(f) = G^*(-f)$ (Hermitian property). We have also that the FT of $g(-t)$ is $G(-f)$. But, if $g(t) = g(-t)$ (g even), then the FT of $g(t)$ is $G(-f) = G(f)$. Therefore, for real and even g, $G(-f) = G^*(-f)$, showing that G is a real function. So the FT of a real and even function is real and even. By a similar argument, one can show that the FT of a real and odd function $[g(t) = -g(-t)]$ is imaginary and odd.

EXAMPLE 7.2 Fourier transform of a truncated cosine wave
Let us find the Fourier transform of a signal

$$y = y_0 \cos 2\pi f_0 t,$$

which starts at time $-T$ and stops at time $+T$, as illustrated in Fig. 7.5. One could imagine this signal as a pure sound of limited duration, such as one of the pulses which make up the Greenwich time signal. Such a signal is, of course, not periodic; it may, perhaps, be thought of as periodic between the finite time limits $-T$ and $+T$ but is certainly not so over the whole time scale. We know that if the value of T were infinity the function would extend over all time, and therefore would be truly periodic, there being only one frequency component at the frequency f_0. (Strictly speaking, of course, there would also be a component at a frequency $-f_0$, because we have chosen in our theory to include negative frequencies.) The problem now is to determine the effect on the frequency spectrum of limiting the duration of the pure tone.

To do this, we must use Fourier transforms. The function $y(t)$ that we wish to transform is characterised mathematically by the following description:

$$y(t) = \begin{cases} y_0 \cos 2\pi f_0 t & \text{for } |t| \leqslant T, \\ 0 & \text{for } |t| > T, \end{cases}$$

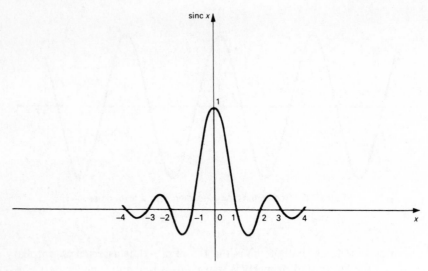

Figure 7.6 Graph of sinc function.

The Fourier transform of $y(t)$ – let us call it $Y(f)$ – is given by (7.15):

$$Y(f) = \int_{-\infty}^{\infty} y(t) \exp(-2\pi ift) \, dt.$$

Since $y(t)$ is zero for t less than $-T$ and greater than $+T$, the limits of the integral now assume these time values. So

$$Y(f) = \int_{-T}^{T} y_0 \cos 2\pi f_0 t \, \exp(-2\pi ift) \, dt$$

$$= \int_{-T}^{T} y_0 \tfrac{1}{2}[\exp 2\pi if_0 t + \exp(-2\pi if_0 t)] \exp(-2\pi ift) \, dt$$

$$= \tfrac{1}{2} y_0 \int_{-T}^{T} \{\exp 2\pi i(f_0 - f)t + \exp[-2\pi i(f_0 + f)t]\} \, dt$$

$$= \frac{y_0}{2} \left\{ \frac{\exp 2\pi i(f_0 - f)t}{2\pi i(f_0 - f)} + \frac{\exp[-2\pi i(f_0 + f)t]}{-2\pi i(f_0 + f)} \right\}_{-T}^{+T},$$

which, after substitution of the limits, becomes

$$\frac{y_0}{2} \left\{ \frac{\exp 2\pi i(f_0 - f)T - \exp[-2\pi i(f_0 - f)T]}{2i} \frac{1}{\pi(f_0 - f)} \right.$$

$$\left. - \frac{\exp[-2\pi i(f_0 + f)T] - \exp 2\pi i(f_0 + f)T}{2i} \frac{1}{\pi(f_0 + f)} \right\}.$$

Since $[\exp i\theta - \exp(-i\theta)]/2i = \sin \theta$, this becomes

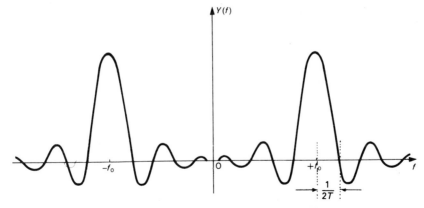

Figure 7.7 Fourier transform of the function shown in Fig. 7.5.

$$Y(f) = \frac{y_0}{2} \left[\frac{\sin\ 2\pi(f_0 - f)T}{\pi(f_0 - f)} + \frac{\sin\ 2\pi(f_0 + f)T}{\pi(f_0 + f)} \right],$$

which, finally, can be put in the form

$$Y(f) = y_0 T \left[\frac{\sin\ 2\pi(f_0 - f)T}{2\pi(f_0 - f)T} + \frac{\sin\ 2\pi(f_0 + f)T}{2\pi(f_0 + f)T} \right].$$

How do we interpret this Fourier transform? First of all, we notice that each of the two terms in the brackets is of the form $\sin \pi x/\pi x$ (sometimes called sinc x) where x in the first case is $2(f_0 - f)T$ and in the second case $2(f_0 + f)T$. The last equation can now be written in the form

$$Y(f) = y_0 T[\text{sinc}\ 2(f_0 - f)T + \text{sinc}\ 2(f_0 + f)T]. \tag{7.16}$$

The graph of sinc x against x is sketched in Fig. 7.6.

A little thought will reveal that the right-hand side of (7.16) contains two functions of this type, the first centred on the frequency f_0 and the second on $-f_0$. The transform is sketched in Fig. 7.7.

The Fourier transform in this example shows a characteristic of non-periodic functions. It is a *band spectrum*; that is, unlike the spectrum in Fig. 7.4, it exists continuously over a range of frequencies. Also, $Y(f)$ is real and even, as befits a transform of the real and even function $y(t)$. ∎

7.4 The uncertainty principle

In example 7.2, the greatest amplitude occurs at the frequency f_0 of the pulse. However, the shortness of duration of the pulse has introduced more frequency components around the frequency f_0. In fact, examination of (7.16) shows that the shorter the total duration, $2T$, of the pulse, the broader will be the sinc function. Conversely, the longer the duration (i.e. the closer $y(t)$ is to the ideal infinitely extending cosine

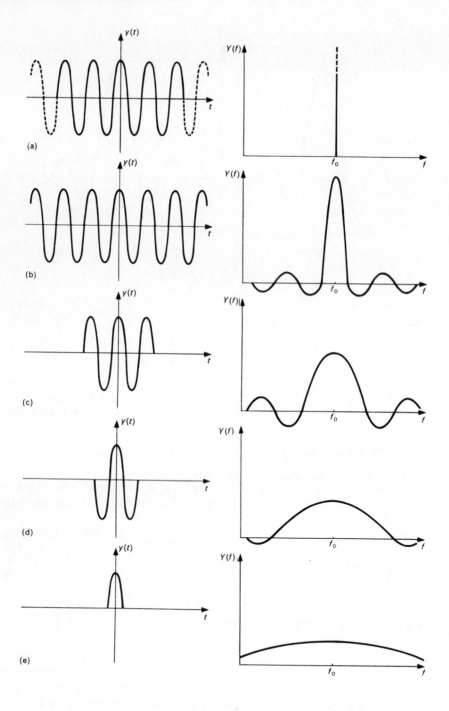

Figure 7.8 Effect on the Fourier transform of termination of signal.

function), the narrower the sinc function becomes. Figure 7.8 illustrates the positive-frequency region of the spectrum of signals, each of frequency f_0 with different times, $2T$, of duration. The infinitely extending signal, illustrated in Fig. 7.8a, produces a 'line' spectrum – that is to say, the sinc function is infinitely narrow, centred on f_0 and infinitely high. (It may, however, be shown that the area under the graph in this case is finite.) As the duration of the signal decreases, Fig. 7.8b, c, d and e shows that the maximum amplitude (which (7.16) shows to be proportional to T) of $Y(f)$ decreases, and the broadness increases. The effect of the finite duration therefore is to 'broaden' the original 'line'. This illustrates an important principle, called the *uncertainty principle*, which states that the product of the duration of the signal and the frequency width of the spectrum of the signal is of the order of unity. In other words, the longer the signal lasts, the narrower is its spectrum and vice versa. The validity of the uncertainty principle can be quite easily demonstrated mathematically for this particular signal. Suppose we define the duration as Δt and the width of the spectrum as Δf. Then clearly $\Delta t = 2T$. However, a little difficulty arises in defining the width Δf of the corresponding frequency spectrum. Since most of the amplitude is in the central peak of the sinc function, let us choose to define Δf as, say, half the spread in frequency of the central peak. The first zero of the sinc function below the frequency f_0 occurs, according to (7.16), when

$$\sin 2\pi(f_0 - f)T = 0,$$

i.e. when $2\pi(f_0 - f)T = \pi$ or when $f_0 - f = 1/(2T)$.

Since we have defined Δf as $f_0 - f$, this means that $\Delta f = 1/(2T)$. Thus

$$\Delta t\, \Delta f = 2T \frac{1}{2T} = 1,$$

which verifies the original assertion. Of course, much depends upon how we choose to define the effective width of the spectrum, and different definitions will yield different values for $\Delta t\, \Delta f$. Mathematical analyses of different signals using a definition of effective width based on criteria of more general applicability yield a value of $1/4\pi$ for $\Delta f\, \Delta t$. So we can take it that a more thorough investigation shows the uncertainty principle to be described by

$$\boxed{\Delta t\, \Delta f \geqslant \frac{1}{4\pi}.} \tag{7.17}$$

However, the important point is that the number on the right-hand side of (7.17) is not zero; that there does exist, as a consequence of the nature of time and frequency, this connection between one and the other.

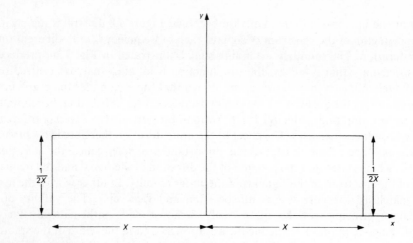

Figure 7.9 The function $y(x) = 0 \quad |x| > X$
$$= 1/2X \quad |x| \leqslant X.$$

7.5 The delta function

In connection with the uncertainty principle and with many other topics related to Fourier transforms, an important concept is that of the *delta function*. The delta function may be considered as describing the spectrum of the cosine function as its duration $2T$ tends to infinity. The nearer this limit of duration is approached, the narrower the spectrum becomes (according to the uncertainty principle) until, when the cosine function is infinitely extended, the spectrum is of infinitesimal width and becomes an example of what is known as a *line spectrum*. The entity describing such a limiting situation is the delta function. However, it turns out that a simpler (and equally valid) approach is to consider it as the limit of a different function which is illustrated in Fig. 7.9. This is a function $y(x)$ which is equal to zero for distances greater than some value X on either side of the origin of x, and to $1/2X$ for distances less than X. In other words,

$$y(x) = \begin{cases} 0 & |x| > X, \\ 1/2X & |x| \leqslant X. \end{cases} \tag{7.18}$$

These values have been chosen so that the integrated area,

$$\int_{-\infty}^{+\infty} y(x) \, dx,$$

is unity, since, clearly, this is equal to the product of the height $1/2X$ of the rectangle and its width $2X$. This integrated area is therefore independent of the value of X.

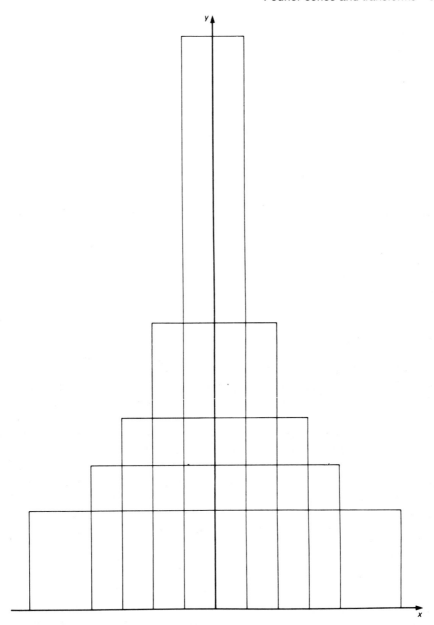

Figure 7.10 The function described by (7.18) for different values of *X*. Each rectangle is of unit area, but the width decreases as *X* decreases. In the limit of *X* = 0, the width is infinitesimal and the height infinite; the integrated area, however, is still unity.

Let us now investigate what happens when we let X tend to zero. The height of the function will increase, and its width will shrink until, in the limit, the width will be infinitesimal and the height infinite. The integrated area, however, will still be unity. Figure 7.10 shows how the limit is approached. We can define the delta function to be the limiting case of $y(x)$ as X tends to zero, and we denote the function by the symbol $\delta(x)$. Thus

$$\delta(x) = \lim_{X \to 0} \ y(x).$$

The symbol $\delta(x)$ should not be confused with δx meaning a small finite increment of x. The brackets after the δ should eliminate this ambiguity. The symbol $\delta(x)$ is due to Dirac who described the delta function as an 'improper function'. It is equal to zero for all x except $x = 0$ at which point it is infinitely large in such a way as to make the integral of $\delta(x)$ over all x equal to unity. It is this singularity which precludes the delta function from being an ordinary function and which therefore provokes Dirac's description. The symbol clearly indicates the position of the singularity, which is that for which the contents of the bracket equals zero. So, for example, the symbol $\delta(x - a)$ is a mathematical description of a delta function of unit integrated area with its singularity at $x - a = 0$, or $x = a$ and, by the same reasoning, $\delta(x)$ represents one with its singularity at $x = 0$.

For certain applications we may wish to specify a delta function, at a particular value of x, of integrated area other than unity; all we need to do in this case, of course, is to multiply the delta function by the appropriate amount. Thus, if we wish to specify a delta function of integrated area 2 with its singularity at $x = -3$, we write $2\delta(x + 3)$. A convenient method of representing delta functions graphically is by drawing a vertical line of height proportional to the integrated area. Some examples are given in Fig. 7.11.

The method of defining the delta function as the limit, as $X \to 0$, of the function depicted in Fig. 7.10 is easy to visualise, and illustrates the idea of a 'spike' of infinitesimal width and infinite height, but with finite area, at the origin. The importance of the function does not reside in the details of its shape, which in this case is rectangular, but in its non-zero integrated area and its infinitesimal width. In fact, we can take limiting cases of a variety of functions to define the delta function. We could start with the so-called *Gaussian* function

$$\sqrt{\frac{a}{\pi}} \exp(-ax^2),$$

which, in the form given, has unit integrated area, and define the delta function as the limit, as $a \to \infty$, of that. Alternatively, we could take the *Lorentzian* function

$$\frac{a}{\pi(a^2 + x^2)}$$

and proceed to the limit as $a \to 0$. In fact, there is an infinite number of suitably chosen functions that we could use.

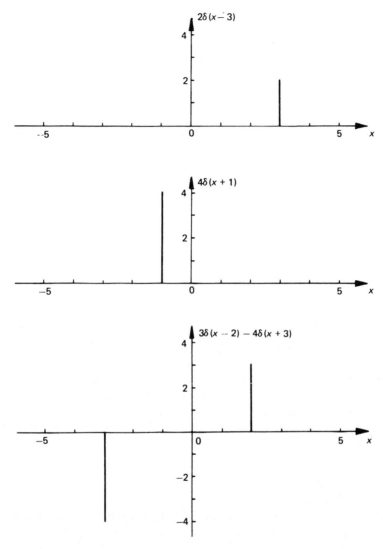

Figure 7.11 Some examples of delta functions and their graphical representation. The heights of the lines representing the delta functions are proportional to their integrated areas. In (c) there are two delta functions on one graph, and the one at $x = -3$ is of magnitude -4.

However, there is a more elegant way of defining the delta function, of general applicability, which does not rely on our proceeding to a limit of some function, such as those above, whose shape has essentially nothing to do with the delta function itself. We consider *any* function, $\phi(x)$, which is finite and continuous, and we define the delta function, $\delta(x)$, as that entity for which

$$\int_{-\infty}^{\infty} \phi(x)\delta(x)\,\mathrm{d}x = \phi(0) \tag{7.19}$$

regardless of our choice of $\phi(x)$. A little thought will reveal that only if $\delta(x)$ is of the form of a spike at the origin with zero value everywhere else can the equation (7.19) be true for *any* $\phi(x)$.

We can derive from (7.19) a result we shall soon need. What is the value of

$$\int_{-\infty}^{\infty} \phi(x)\delta(x - x_0)\,\mathrm{d}x$$

where x_0 is a constant? Putting $x - x_0 = u$, we see that the expression becomes

$$\int_{-\infty}^{\infty} \phi(u + x_0)\delta(u)\,\mathrm{d}u$$

which, by (7.19), is equal to $\phi(x_0)$. Thus

$$\int_{-\infty}^{\infty} \phi(x)\delta(x - x_0)\,\mathrm{d}x = \phi(x_0). \tag{7.20}$$

Let us now see how the delta function concept works in a simple, but very important, case. We suspect, from our previous discussion on the uncertainty principle, that the Fourier transform of $\cos 2\pi f_0 t$ (where the function has infinite duration for positive and negative time) consists of a delta function at the value f_0 along the f-axis and another of equal magnitude with singularity at the value $-f_0$. We do not know what the magnitude is; let us therefore call it A. Now if we tackled the problem directly by applying (7.15) we would see that the Fourier transform of $\cos 2\pi f_0 t$ is

$$\int_{-\infty}^{\infty} \cos 2\pi f_0 t \, \exp(-2\pi i f t)\,\mathrm{d}t,$$

but we cannot go further than this because the integral cannot be evaluated. However, assuming, as we have done, that the Fourier transform is

$$A[\delta(f - f_0) + \delta(f + f_0)], \tag{7.21}$$

we can transform back into time from frequency and see if we obtain the function $\cos 2\pi f_0 t$. To do this we use (7.14), substituting the expression (7.21) for $G(f)$. So $g(t)$ is given by

$$g(t) = \int_{-\infty}^{\infty} A[\delta(f - f_0) + \delta(f + f_0)] \exp 2\pi i f t \, \mathrm{d}f,$$

$$= A \int_{-\infty}^{\infty} \delta(f - f_0) \exp 2\pi i f t \, \mathrm{d}f + A \int_{-\infty}^{\infty} \delta(f + f_0) \exp 2\pi i f t \, \mathrm{d}f.$$

If we consider the first integral, we see that it is of the form (7.20) with f and f_0 for x and x_0, and $\exp(2\pi ift)$ for $\phi(x)$. The value of the integral is therefore $A\exp(2\pi if_0 t)$. Similarly, the second integral is $A\exp(-2\pi if_0 t)$. Thus the full equation now becomes

$$g(t) = A[\exp(2\pi if_0 t) + \exp(-2\pi if_0 t)].$$

Furthermore, since $\cos\theta = \frac{1}{2}[\exp(i\theta) + \exp(-i\theta)]$,

$$g(t) = 2A\cos 2\pi f_0 t.$$

Putting $A = \frac{1}{2}$, we see that $g(t)$ is, indeed, $\cos 2\pi f_0 t$. Thus the Fourier transform of $\cos 2\pi f_0 t$ is

$$\frac{1}{2}[\delta(f + f_0) + \delta(f - f_0)]. \tag{7.22}$$

The transform pair is illustrated in Fig. 7.12.

An interesting fact emerges when we make f_0 zero. The function $\cos 2\pi f_0 t$ becomes unity over all time and its Fourier transform becomes

$$\frac{1}{2}\delta(f - 0) + \frac{1}{2}\delta(f + 0) = \delta(f),$$

this is, a delta function at the origin. In other words, a steady d.c. voltage has a spectrum entirely at the zero of frequency. Furthermore, by using the symmetry of t and f in the transform relationship as described in an earlier section, we see that the

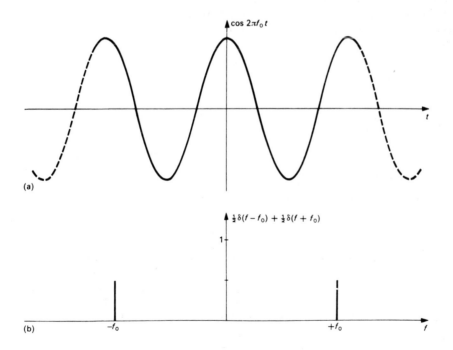

Figure 7.12 (a) The function $\cos 2\pi f_0 t$, and (b) its Fourier transform.

Fourier transform of $\delta(t)$ is unity for all frequencies. In other words, the spectrum of a sharp pulse contains equal amplitude at all frequencies. Indeed, we can derive this very easily. The transform of $\delta(t)$ is, by (7.15),

$$\int_{-\infty}^{\infty} \delta(t) \exp(-2\pi ift)\, dt$$

which is of the form (7.19) with $\phi(t) = \exp(-2\pi ift)$. Since $\phi(0) = 1$, we see immediately that the transform of $\delta(t)$ is unity.

We now see more clearly the nature of the delta function. It is not a description of anything that could exist in nature but it is a very useful idealisation. No signal of the description $\cos 2\pi f_0 t$ can be produced by any apparatus, since such apparatus would need to have been made (and switched on) an infinite time ago, and moreover would have to continue to operate for an infinite time to come. Consequently, there can be no spectra consisting of delta functions. Nevertheless, in practice we can get very close to the line-spectrum situation which makes the concept important.

The delta function is by no means limited to discussions on Fourier transforms; it is unconsciously built into our thinking at an early age. The particle of classical mechanics is introduced as a body having a finite mass but no size. It is the limiting case of finite-sized body of mass m as the size tends to zero, the density meanwhile tending to infinity in such a way as to keep the total mass m a constant. In other words, the particle is none other than a three-dimensional version of a delta function of magnitude m. In the same way we can treat point electric charges, and many other physical concepts, as delta functions.

Problems

1.   For the trigonometrical Fourier-series representation of a periodic function, show that the Fourier coefficients S_n are all zero if the function is even (i.e. $g(t) = g(-t)$) and that the coefficients C_n are all zero if the function is odd (i.e. $g(t) = -g(-t)$). In each of these cases, what property is possessed by G_n, the complex Fourier coefficients of the exponential representation?

2.   A certain periodic function of period T is equal to $h\cos(2\pi f_0 t)$ for $-T/4 \leqslant t \leqslant +T/4$ and zero for the remainder of the period. Sketch the function over a few periods. (It is the form of signal obtained when a cosine signal is passed through a half-wave rectifier. Such a rectifier gives out the undistorted signal when the input has a positive value and zero when it has a negative or zero value.) Perform an exponential Fourier analysis of the function and show that the complex Fourier coefficients G_n are

$$G_n = \frac{h}{4}\left[\text{sinc}\left(\frac{1-n}{2}\right) + \text{sinc}\left(\frac{1+n}{2}\right)\right]$$

or, alternatively,

$$G_n = \frac{h}{\pi(1-n^2)} \cos\left(\frac{\pi n}{2}\right).$$

Note that all the coefficients for odd n are missing except G_1 and G_{-1}. While the exponential Fourier series is easier to handle analytically, the cosine series is essential for a computer synthesis. Convert the G_n, therefore, into amplitudes A_n for the cosine series using $|G_n| = A_n/2$. Work out explicitly the coefficients A_n for n up to 10 and synthesise the function on the computer using demonstration No. 7 on the available disk.

3. A function $g(t)$ of period T is defined between $t = 0$ and $t = T$ as follows:

$$g(t) = h \text{ for } 0 < t \leqslant T/2,$$

$$g(t) = -h \text{ for } T/2 < t \leqslant T.$$

Perform an exponential Fourier analysis of $g(t)$ and show that the complex Fourier coefficients G_n are

$$G_n = i(h/\pi n)[\cos(\pi n) - 1],$$

Plot the coefficients for the first 12 positive harmonics graphically. Convert your results into trigonometrical form as for the previous problem and synthesise the function on the computer.

4. Suppose that you know the complex Fourier coefficients G_n of a particular periodic function $g(t)$ of period T. If the coefficients of the shifted version, $g(t - t_0)$, of the same signal are H_n, show that

$$H_n = \exp(-2\pi i t_0 n f_0)G_n,$$

so that you do not actually have to perform a Fourier analysis of the shifted version. (This is an example of the so-called *shift theorem*, which is treated in some detail in the context of Fourier transforms in the next chapter.)

If the shift t_0 is an integral number of periods, what is the relationship between H_n and G_n? Comment on your answer.

5. Show that the FT, $X(f)$, of any real function $x(t)$ is Hermitian, that is, $X(f) = X^*(-f)$. Hence show that the amplitude spectrum $|X(f)|$ of any real function is even. (This is an important result in FT theory; if you know that $x(t)$ is a real function (they nearly always are in physical applications) then only the amplitude spectrum at positive frequencies need be calculated and interpreted.)

What is the relationship of the phase spectrum at positive frequencies to that at negative frequencies?

6. If

$$g(t) = \exp(-|t|/T)/2T,$$

show that

$$G(f) = 1/(1 + 4\pi^2 f^2 T^2).$$

Sketch $g(t)$ and $G(f)$. Without performing any integration, deduce the FT of the function $1/(1 + t^2/\tau^2)$.

7. A burst of 10 cycles from a 5 kHz sinusoidal oscillator is fed into an amplifier. What frequency bandwidth must the amplifier possess to give a reasonably undistorted output?

8. For a real, non-zero constant a, show that $\delta(ax)$ may be interpreted as a delta function which has its singularity at the origin of x and has magnitude $1/|a|$. Hence, show that $\delta(x)$ behaves as though it were an even function.

9. An infinite series of unit-magnitude delta functions with their singularities at $t = \ldots -2T$, $-T, 0, T, 2T, \ldots$ may be represented as

$$g(t) = \sum_{m=-\infty}^{\infty} \delta(t - mT).$$

When plotted after the manner of Fig. 7.11, $g(t)$ has the appearance of a *comb*, and this is the name by which $g(t)$ is usually known. It is periodic in t with period T. Analyse it into an exponential Fourier series. For this purpose, note that the defining integral for the delta function

$$\int_{-\infty}^{\infty} \phi(t)\delta(t)\, dt = \phi(0)$$

may be re-expressed as

$$\int_{-T/2}^{T/2} \phi(t)\delta(t)\, dt = \phi(0)$$

since the integral outside the region $-T/2 \leqslant t \leqslant T/2$ is zero.

Show that all the coefficients G_n are equal and try to synthesise the comb with the disk using as many coefficients as you can. Note that the approximation to the comb gets better the larger the number of coefficients you include. (The comb is an important function which is considered from a Fourier-transform, rather than Fourier-series, point of view in the next chapter.)

8 Further topics in Fourier theory

8.1 Some theorems

It may be thought at first that, in order to obtain a Fourier transform (FT) of interest, one would have to perform some integration, in view of the form of (7.15). In fact, however, it is often possible to transform a function without integrating, because of the existence of certain theorems to be described below. As we shall see, the theorems allow one to see the relationship between the FT of a function and that of others which are related to it in certain ways. From a few basic functions (whose FTs may have to be obtained by integration) we can build up a 'library' – without further integration – of FTs, with the use of these theorems.

8.1.1 The addition theorem

This merely states that the FT of the sum of two functions is the sum of the individual FTs. Suppose we know the FTs $G_1(f)$ and $G_2(f)$ of the functions $g_1(t)$ and $g_2(t)$ respectively. Then the FT of

$$h(t) = g_1(t) + g_2(t)$$

is

$$H(f) = G_1(f) + G_2(f). \tag{8.1}$$

The proof, which is trivial, will not be given.

8.1.2 The shift theorem

Suppose the FT of $g(t)$ is $G(f)$. Then, we ask, what is the FT of $g(t - t_0)$ (where t_0 is a constant) in terms of $G(f)$? The function $g(t - t_0)$ is, of course, the function $g(t)$ shifted bodily an amount t_0 in the direction of positive t. To obtain its FT, say $H(f)$, we start with (7.15), except that now we wish to transform $g(t - t_0)$ instead of $g(t)$.

Therefore

$$H(f) = \int_{-\infty}^{\infty} g(t - t_0) \exp(-2\pi i f t) \, dt.$$

We proceed by putting $u = t - t_0$, so that $du = dt$. The integral then becomes

$$H(f) = \int_{-\infty}^{\infty} g(u) \exp[-2\pi i f(u + t_0)] \, du.$$

The factor $\exp(-2\pi i f t_0)$ is a constant which may be taken outside the integral. So

$$H(f) = \exp(-2\pi i f t_0) \int_{-\infty}^{\infty} g(u) \exp(-2\pi i f u) \, du.$$

Therefore, from (7.15),

$$H(f) = \exp(-2\pi i f t_0) G(f). \tag{8.2}$$

This is the so-called shift theorem; if a function of t is translated along the time axis (shifted), then its FT is multiplied by $\exp(-2\pi i f t_0)$. It will be noted that this latter factor has a modulus of unity. So $|H(f)| = |G(f)|$; that is, if we shift a function of t, then there is no change whatsoever in its amplitude spectrum. That this is so is not difficult to see. If a sound is played on two separate occasions (so that if the signal is $g(t)$ on the first occasion then it is $g(t - t_0)$ on the second occasion t_0 later) then it is clear that its amplitude spectrum must be the same on both occasions.

The shift theorem provides us with general information about the spectrum of a sound together with an echo of itself. Suppose, for simplicity, that the original sound is $g[t + (t_0/2)]$ and that its echo is an exact replica of the original except that it is delayed by a time t_0, so that it is $g[t - (t_0/2)]$. Then the complete signal, say, $h(t)$, is

$$h(t) = g\left(t + \frac{t_0}{2}\right) + g\left(t - \frac{t_0}{2}\right).$$

By the addition (8.1) and shift (8.2) theorems,

$$H(f) = [\exp(2\pi i f t_0/2) + \exp(-2\pi i f t_0/2)]G(f),$$

where $G(f)$ is the FT of $g(t)$.

Since $\cos\theta = [\exp(i\theta) + \exp(-i\theta)]/2$, we have

$$H(f) = 2\cos(\pi f t_0)G(f).$$

So the spectrum is that of $g(t)$, multiplied by a cosine function which causes certain regularly spaced frequencies, namely $1/2t_0, 3/2t_0, 5/2t_0, (2n + 1)/2t_0, \ldots$, to be absent.

8.1.3 The similarity theorem

This theorem concerns the relationship between the FT of a function $g(t)$ and that of $g(at)$, where a is a constant. We will, for the moment, think of a as positive, but

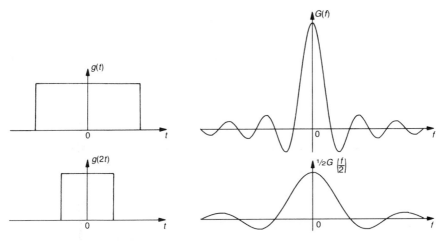

Figure 8.1 Illustration of the similarity theorem. The left-hand diagrams show the relationship between $g(t)$ and $g(2t)$; the right-hand diagrams are the FTs of those on the left.

will remove this restriction later. To see the relationship between $g(t)$ and $g(at)$, let us fix our ideas by supposing a to be 2. The function $g(2t)$ plotted against $2t$ will look exactly the same as $g(t)$ plotted against t. So if $g(2t)$ is plotted against t instead of $2t$, it will appear to be 'squashed', by a factor of 2, along the t-axis. In other words, any given feature of $g(2t)$ will occupy half as much time as the same feature in $g(t)$. In still other words, a sound signal $g(t)$ played by a tape recorder at normal speed will become $g(2t)$ if the tape recorder is played at double the speed. The relationship between $g(t)$ and $g(2t)$ is displayed in the left-hand side of Fig. 8.1.

If the FT of $g(t)$ is $G(f)$, what is the FT of $g(at)$? We start, again, with (7.15) and suppose that the FT of $g(at)$ is $H(f)$. Then

$$H(f) = \int_{-\infty}^{\infty} g(at) \exp(-2\pi i f t)\, \mathrm{d}t.$$

We put $u = at$, so that $\mathrm{d}u = a\, \mathrm{d}t$. Thus

$$H(f) = \int_{-\infty}^{\infty} g(u) \exp(-2\pi i f u/a)\, \mathrm{d}u/a$$

$$= \frac{1}{a} \int_{-\infty}^{\infty} g(u) \exp\left[-2\pi i\left(\frac{f}{a}\right)u\right] \mathrm{d}u. \qquad (8.3)$$

Now, the integral in (8.3) is exactly of the form of (7.15), except that t in (7.15) is replaced by u (which does not matter since it disappears on application of the limits of the integral) and f is replaced by f/a. So (8.3) becomes

$$H(f) = \frac{1}{a} G\left(\frac{f}{a}\right).$$

This is the so-called similarity theorem. Before we interpret, let it be mentioned that if *a* is negative, an analysis identical to the above would yield

$$H(f) = -\frac{1}{a} G\left(\frac{f}{a}\right).$$

Thus two results may be subsumed, for *any* real, non-zero *a*, into a single result

$$H(f) = \frac{1}{|a|} G\left(\frac{f}{a}\right). \tag{8.4}$$

We interpret the similarity theorem as follows. If the original signal $g(t)$ is 'squashed' by a factor of *a* to produce $g(at)$, then its FT, $G(f)$, is 'stretched' in frequency space to produce $(1/|a|)G(f/a)$. (The overall magnitude of the FT is also changed by a factor of $|a|$, but this is not generally of as much interest as the 'stretching'.) In particular, if $a = 2$, then the FT of $g(2t)$ is $\frac{1}{2}G(f/2)$, the $f/2$ in the argument of G showing that the FT of $g(t)$ has been 'stretched' by a factor of 2 to produce the FT of $g(2t)$.

In other words, if a tape recorder gives out the signal $g(2t)$, all the features in its spectrum will be at twice the frequencies that they would be at if the signal were $g(t)$.

A particular case of the similarity theorem is given by $a = -1$, where we see immediately from (8.4) that the FT of $g(-t)$ is $G(-f)$. So reversal of $g(t)$ implies reversal of $G(f)$, and, since the amplitude spectrum of a real function $g(t)$ is even, there is no change in the amplitude spectrum on reversal of $g(t)$. In other words, the amplitudes at various frequencies are the same whether the tape is played backwards or forwards.

The similarity theorem is clearly relevant to the uncertainty principle discussed in the last chapter. If the function $g(t)$ is of finite width, then it is clear that the width of $g(t)$ (*a* assumed positive) decreases as *a* increases. By the same token, that of its FT, $(1/a)G(f/a)$ increases as *a* increases. With appropriate definition of width, the details of which we will not go into, it is possible to show that the product of the widths of $g(at)$ and $(1/a)G(f/a)$ is a finite number independent of *a*, consistent with the uncertainty principle (7.17).

8.1.4 The derivative theorem

As its name implies, the derivative theorem connects the FT of $dg(t)/dt$ with that of $g(t)$. Now, the inverse transform of $G(f)$ is

$$g(t) = \int_{-\infty}^{\infty} G(f) \exp(2\pi i f t) \, df. \tag{7.14}$$

Differentiating both sides with respect to t, we have

$$\frac{dg(t)}{dt} = \int_{-\infty}^{\infty} [2\pi i f G(f)] \exp(2\pi i f t) \, df,$$

an equation of the same form as (7.14). Evidently, then, the contents of the square bracket must be the FT of $dg(t)/dt$. In other words, if $G(f)$ is the FT of $g(t)$, then that of $dg(t)/dt$ is $2\pi i f G(f)$.

This is the *derivative theorem*, that if a function of t is differentiated, then its FT is multiplied by $2\pi i f$. By a simple extension of the above argument, we can see that if a function of t is differentiated n times, then its FT is mutiplied by $(2\pi i f)^n$. There are many obvious uses of the theorem for generating a variety of FTs of functions which are the derivatives with respect to time of a function whose FT is known. In addition, however, the theorem is particularly useful in illuminating certain aspects of linear differential equations.

EXAMPLE 8.1 Fourier-transform analysis of an *RLC circuit*

Let us consider, as an example, the electrical circuit depicted in Fig. 3.8b. Here, a time-varying voltage is applied to an *RLC* series circuit. The voltage need not be sinusoidal; indeed, it can be any suitable signal, for example the output from a microphone or tape recorder. Let us call it $v(t)$. As a result of the application of $v(t)$, a time-varying current, $i(t)$, will flow in the circuit. What is the relationship between $v(t)$ and $i(t)$?

Applying (3.30), and bearing in mind that $i = dq/dt$, we have

$$v(t) = Ri(t) + L\frac{di(t)}{dt} + \frac{q(t)}{C}.$$

Differentiation with respect to t yields

$$\frac{dv(t)}{dt} = R\frac{di(t)}{dt} + L\frac{d^2i(t)}{dt^2} + \frac{i(t)}{C}. \tag{8.5}$$

The technique we will now adopt is to obtain the FT of (8.5) which may be transformed term by term by the addition theorem (8.1). We let the FTs of $v(t)$ and $i(t)$ respectively be $V(f)$ and $I(f)$. Then by the derivative theorem, the FTs of $dv(t)/dt$, $di(t)/dt$ and $d^2i(t)/dt^2$ are, respectively, $2\pi i f V(f)$, $2\pi i f I(f)$ and $(2\pi i f)^2 I(f)$. So the FT of (8.5) is

$$2\pi i f V(f) = 2\pi i f R I(f) + (2\pi i f)^2 L I(f) + \frac{I(f)}{C}. \tag{8.6}$$

We see immediately that the process of Fourier transformation of (8.5) has converted that equation from a differential equation into an ordinary algebraic one. Dividing (8.6) throughout by $2\pi i f$ we obtain

$$V(f) = \left[R + 2\pi i f L + \frac{1}{2\pi i f C}\right] I(f) \tag{8.7}$$

which is an equation well known to those readers who are familiar with alternating-current theory. In fact the Fourier-transform approach is essentially nothing more than a generalisation of phasor methods applied to circuits fed with sinusoidally varying voltages. The point of the generalisation lies, of course, in the fact that *any* time-varying voltage, $v(t)$, may be dealt with. To find $i(t)$ for a given $v(t)$, we solve (8.7) for $I(f)$ and then perform an inverse FT (7.14) on the resulting equation. For a general $v(t)$ this can be quite tricky, and we will not go into details here. ∎

8.2 Convolutions

Convolution is a certain mathematical operation involving two (or, sometimes more) functions. The convolution $f(x)*g(x)$ of two functions $f(x)$ and $g(x)$ is defined as

$$f(x)*g(x) = \int_{-\infty}^{\infty} f(u)g(x-u)\,\mathrm{d}u \tag{8.8}$$

where u is a dummy variable in x space. The first function is considered as being plotted out as a function of u, and the second is reflected about the vertical axis and displaced a distance x. These two functions are then multiplied, and the value of the convolution for the particular value of x chosen is the integrated area of the product. By choosing, in turn, different values of x, we may build up a picture of the convolution as a function of x.

If the functions are of finite width in x space it is usually true that the convolution is broader than either $f(x)$ or $g(x)$. For this reason, convolution is sometimes called a 'smearing' of one function by the other. It turns out, in many cases, that the degradation of information in the form of a signal $f(x)$ that takes place in the passage of the signal through a channel may be expressed as the convolution of $f(x)$ with a function, say $g(x)$, which is a property of the channel.

In this section, we will study some of the properties of convolutions, but will leave a discussion of their relevance to Fourier theory until the next section.

Firstly, we will prove the surprising result that $f(x)*g(x) = g(x)*f(x)$, that is, that convolution is commutative. The right-hand side of (8.8) looks, at first glance, to be alarmingly asymmetrical in f and g. However, if we put $v = x - u$, and remember that x is a constant in this integral, then (8.8) becomes

$$f(x)*g(x) = \int_{+\infty}^{-\infty} f(x-v)g(v)(-\mathrm{d}v)$$

$$= \int_{-\infty}^{\infty} f(x-v)g(v)\,\mathrm{d}v = \int_{-\infty}^{\infty} g(v)f(x-v)\,\mathrm{d}x$$

$$= g(x)*f(x),$$

which proves the commutativity of f and g with respect to convolution.

We next state that convolution is *associative*, i.e. that

$$[f(x)*g(x)]*h(x) = f(x)*[g(x)*h(x)], \tag{8.9}$$

so that the convolution of three functions may be expressed as $f(x)*g(x)*h(x)$. The proof of this is delayed till the next section. Clearly, any number of functions to be convolved may be strung together with asterisks and (in view of the commutativity discussed above) may be written in any order.

A function may be convolved with one or more delta functions, and two or more delta functions may be convolved. Let us consider the simplest case, that of $\delta(x)*f(x)$. It will be useful to recall (7.19):

$$\int_{-\infty}^{\infty} \phi(u)\delta(u)\ du = \phi(0) \qquad [7.19]$$

where $\phi(u)$ is *any* finite and continuous function, and we have (for our later convenience) substituted u for x. Now

$$\delta(x)*f(x) = \int_{-\infty}^{\infty} \delta(u)f(x-u)\ du$$

by definition (8.8). This is exactly of the form (7.19) with $f(x-u)$ substituted for $\phi(u)$. So, by (7.19)

$$\boxed{\delta(x)*f(x) = f(x).} \qquad (8.10)$$

In other words, convolution with a delta function whose singularity is at the origin of x leaves the original function $f(x)$ unaltered; thus $\delta(x)$ is to convolution what unity is to multiplication.

What, then, is the value of the convolution of $f(x)$ with a delta function, $\delta(x-x_0)$, situated at the point $x = x_0$? By definition (8.8) it is

$$\delta(x-x_0)*f(x) = \int_{-\infty}^{\infty} \delta(u-x_0)f(x-u)\ du.$$

We put $u - x_0 = v$, and the convolution becomes

$$\int_{-\infty}^{\infty} \delta(v)f(x-x_0-v)\ dv$$

which, by (7.19) is $f(x - x_0)$. So

$$\delta(x-x_0)*f(x) = f(x-x_0). \qquad (8.11)$$

We see, then, that convolution with a delta function is equivalent to shifting the origin of $f(x)$ to the point x_0 where the singularity of the delta function is situated, producing $f(x - x_0)$.

In our discussion of the shift theorem in section 8.1.2 we considered a signal $g[t + (t_0/2)]$ and its echo t_0 later, $g[t - (t_0/2)]$ giving a total signal

$$h(t) = g\left(t + \frac{t_0}{2}\right) + g\left(t - \frac{t_0}{2}\right).$$

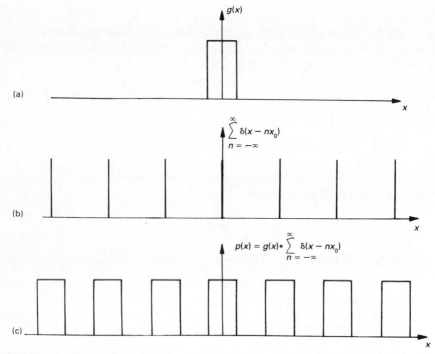

Figure 8.2 Illustration of the convolution representation of a periodic function: (a) is the basic motif of the periodic function, (b) is a series of equally spaced unit-integrated area delta functions and (c) is the convolution of (a) and (b), producing the final periodic function.

From the above result (8.11) concerning convolutions of functions with delta functions we can now see that $h(t)$ may be represented by a convolution,

$$h(t) = g(t)*\left[\delta\left(t + \frac{t_0}{2} \right) + \delta\left(t - \frac{t_0}{2} \right) \right]. \tag{8.12}$$

So we have a representation of $h(t)$ which separates the form of the signal $g(t)$ from details of the echo. This is a simple example of the way in which the convolution representation can separate sharply the information being conveyed $[g(t)]$ from the degradation being performed on it (in this case $\delta[t + (t_0/2)] + \delta[t - (t_0/2)]$) by a system. The representation (8.12) has a further significance as we shall see in the next section.

Finally, in this section, we turn our attention to the convolution representation of periodic functions.

We can consider a periodic function as consisting of a motif laid out at periodic intervals along the x-axis. If the motif (the contents of one period) is $g(x)$, and identical motifs are to be situated at distance x_0 apart, then the periodic function, $p(x)$, may be represented by the function

$$p(x) = \sum_{n=-\infty}^{\infty} g(x - nx_0).$$

But, by (8.11)

$$g(x - nx_0) = g(x)*\delta(x - nx_0).$$

So

$$p(x) = \sum_{n=-\infty}^{\infty} g(x)*\delta(x - nx_0),$$

and, since $g(x)$ is common to all terms in the summation,

$$p(x) = g(x)*\left[\sum_{n=-\infty}^{\infty} \delta(x - nx_0) \right]. \tag{8.13}$$

Here, we have, in the convolution representation of a periodic function, a separation made between the details of the motif [$g(x)$] and those of the periodicity [$\sum_n \delta(x - nx_0)$]. For a given periodicity, only the first term in the convolution gives any information concerning what it is that is being repeated. It is as if two aspects of a periodic function have been 'factorised' into two separate 'factors'. An illustration is given in Fig. 8.2.

A perfect crystal, which is a periodic arrangement of a motif in three dimensions, may be represented in a way similar to (8.13). Here, we are able to represent the crystal as the convolution of the contents of the so-called *unit cell* (the motif) with a triply periodic set of delta functions, called the *lattice*. Crystallographers find this separation very useful, and when they are investigating a crystal structure, usually determine the details of the lattice dimensions and symmetry before concentrating on those of the motif.

8.3 The convolution theorem

In the last section, we dealt with the nature of the convolution, without discussing its role in Fourier theory. We now proceed with this discussion and start by asking the question: what is the Fourier transform of the convolution of a pair of functions? In other words, what is the value, $H(f)$, of the integral

$$H(f) = \int_{-\infty}^{\infty} [g_1(t)*g_2(t)]\, \exp(-2\pi ift)\, dt?$$

We first expand the integral according to (8.8) and obtain

$$H(f) = \int_{-\infty}^{\infty} \int_{-\infty}^{\infty} g_1(u)g_2(t - u)\, \exp(-2\pi ift)\, du\, dt$$

$$= \int_{-\infty}^{\infty} g_1(u) \left[\int_{-\infty}^{\infty} g_2(t - u)\, \exp(-2\pi ift)\, dt \right] du.$$

We work out the inner integral first, and put $y = t - u$. Noting that u is a constant in this integral because we are integrating with respect to t, we have $dy = dt$. So

$$H(f) = \int_{-\infty}^{\infty} g_1(u) \left[\int_{-\infty}^{\infty} g_2(y) \exp[-2\pi i f(y+u)] \, dy \right] du$$

$$= \int_{-\infty}^{\infty} g_1(u) \exp(-2\pi i f u) \left[\int_{-\infty}^{\infty} g_2(y) \exp(-2\pi i f y) \, dy \right] du.$$

But the inner integral is the FT, $G_2(f)$, of $g_2(y)$ by (7.15). So

$$H(f) = \int_{-\infty}^{\infty} g_1(u) \exp(-2\pi i f u) G_2(f) \, du$$

$$= G_2(f) \left[\int_{-\infty}^{\infty} g_1(u) \exp(-2\pi i f u) \, du \right].$$

Now the contents of the square bracket are, by (7.15), the FT, $G_1(f)$, of $g_1(u)$. So

$$H(f) = G_1(f)G_2(f). \tag{8.14}$$

Equation (8.14) is a statement of the *convolution theorem*, that the FT of a convolution of two functions is the product of the individual transforms. If we convolve two functions in time, then we mutiply the two transforms in frequency space.

The equation (8.14) can easily be extended to give the FT of

$$g_1(t)*g_2(t)* \ldots *g_n(t)$$

(i.e. the convolution of n functions) which is

$$G_1(f)G_2(f) \ldots G_n(f). \tag{8.15}$$

The commutativity and associativity of the convolution operation, discussed in the last section, is now very obvious. Since in (8.14) $G_1(f)G_2(f)$ is equal to $G_2(f)G_1(f)$, it follows that $g_1(t)*g_2(t) = g_2(t)*g_1(t)$. Similarly, since, in (8.15),

$$[G_1(f)G_2(f)]G_3(f) = G_1(f)[G_2(f)G_3(f)]$$

then we must have

$$[g_1(t)*g_2(t)]*g_3(t) = g_1(t)*[g_2(t)*g_3(t)].$$

There is another form of the convolution theorem that can be useful. In view of the great similarity in form of the two equations (7.14) and (7.15), we may start in frequency space instead of time, and ask what function of time would, when Fourier-transformed, give a convolution of two functions in frequency space. By an analysis identical to the above except that we start in frequency space and proceed to time, we can show that the inverse FT of $G_1(f)*G_2(f)$ is $g_1(t)g_2(t)$, where $g_1(t)$ and $g_2(t)$ are the inverse FTs of $G_1(f)$ and $G_2(f)$ respectively. So the convolution theorem displays a symmetry between time and frequency spaces. If we convolve two functions in time space, we multiply their FTs in frequency space; if we multiply two functions in time space, we convolve their FTs in frequency space.

Let us take a simple example of the application of the convolution theorem. For this, we will need the FT of a delta function situated at the time t_0. The expression for this delta function is of course, $\delta(t - t_0)$. Its FT is

$$\int_{-\infty}^{\infty} \delta(t - t_0) \exp(-2\pi i f t) \, dt,$$

by (7.15), which, by virtue of (7.20), becomes

$$\exp(-2\pi i f t_0). \tag{8.16}$$

Now the example that we take is the calculation of the FT of a signal $g(t - t_0)$, which, of course, is a shifted version of a signal $g(t)$. Since, by (8.11), $g(t - t_0) = \delta(t - t_0) * g(t)$, its FT is, by the convolution theorem (8.14), the product of the FTs of each function in the convolution. Because (by (8.16)) the FT of $\delta(t - t_0)$ is $\exp(-2\pi i f t_0)$, we have that the FT of $g(t - t_0)$ is $\exp(-2\pi i f t_0)G(f)$, a result identical to the shift theorem (8.2). We see, therefore, that the shift theorem is merely a special case of the convolution theorem in which one of the two functions convolved is the $\delta(t - t_0)$ which supplies the information concerning the amount (t_0) by which the function $g(t)$ has been shifted.

Returning to expression (8.16), and putting $t_0 = 0$, we see that the FT of a delta function at the origin of time, $\delta(t)$, is 1. So an infinitesimally wide pulse has an FT which is a constant at all frequencies. Therefore, in order to reproduce such a sound signal we would need an amplifier which would amplify equally all frequencies. Neither such a sound signal nor such an amplifier could exist in practice, and the situation described represents an idealisation.

Since any signal $g(t)$ can be represented as $\delta(t) * g(t)$, its FT is, by the convolution theorem (8.14), $1 \times G(f) = G(f)$, giving us an obvious truth. Trivial though this example may be, it illustrates, once again, that a delta function at the origin is to convolution what unity is to multiplication.

We turn our attention now to the convolution representation of a signal with its echo

$$h(t) = g(t) * \left[\delta\left(t + \frac{t_0}{2}\right) + \delta\left(t - \frac{t_0}{2}\right) \right]. \tag{8.12}$$

The FT of the contents of the square bracket is, by (8.16)

$$\exp\left(+2\pi i f \frac{t_0}{2}\right) + \exp\left(-2\pi i f \frac{t_0}{2}\right) = 2 \cos(\pi f t_0).$$

So, by the convolution theorem (8.14), the FT of equation (8.12) is

$$H(f) = 2G(f) \cos(\pi f t_0),$$

a result which we obtained earlier from the shift theorem.

Finally in this section, we consider the FT of a periodic function. We have already seen that a periodic function $p(t)$ may be represented as a convolution of a motif $g(t)$ with a series of delta functions $\sum_{n=-\infty}^{\infty} \delta(t - nT)$ spaced equally apart at the periodic time T:

$$p(t) = g(t)* \sum_{n=-\infty}^{\infty} \delta(t - nT).$$

So $P(f)$, the FT of $p(t)$, is, by convolution theorem (8.14), equal to the product of $G(f)$ and the FT of the sum of the delta functions. We therefore need to find the latter FT. Unfortunately, this is not a particularly easy problem, and is outside the scope of this book. The result, however, is simple and will now be quoted. The FT of $\sum_{n=-\infty}^{\infty} \delta(t - nT)$ is

$$\frac{1}{T} \sum_{m=-\infty}^{\infty} \delta\left(f - \frac{m}{T}\right).$$

So the FT of a sequence of identical equally spaced delta functions is itself a sequence of identical equally spaced delta functions. Such a sequence is often called a *comb*, by reason of its appearance when plotted (see Fig. 8.2b). The FT of a comb is therefore another comb. The 'teeth' of the comb in t space are a time T apart, and those of the comb in f space are $1/T$ apart. So the coarser the comb in t space is, the finer is that in f space, and *vice versa*. A proof of this result is given *in The Fourier Transform and its Applications* by R. Bracewell, McGraw-Hill, 2nd edition, 1978.

The FT of $p(t)$ is shown pictorially in Fig. 8.3. Here we see in (c) the periodic function $p(t)$ expressed as the convolution of $g(t)$ (a) with the comb (b). The latter defines the periodicity of the function $p(t)$. On the right-hand side of the figure we see the FTs of the functions on the left-hand side. $G(f)$ is the FT (d) of the motif; (e) is the comb which is the FT of the comb (b) on the left-hand side. The final $P(f)$ is depicted in (f). We see in (f) that the positions of the delta functions define the frequencies at which components are situated, and the amplitudes of those components are determined by the local value of $G(f)$, the FT of the basic motif.

In the first sections of this and the previous chapter, we have made a tour of Fourier theory which has come full circle. We started with periodic signals and found that their spectra consisted only of components at discrete frequencies which were integral multiples of the fundamental. Then we discussed non-periodic signals and discovered that their spectra generally contained finite amplitudes at all frequencies. We saw, from the Fourier transform, the remarkable symmetry that exists between time and frequency spaces culminating in the uncertainty principle (7.17). Finally, we turned again to periodic functions and found that they can be thought of in Fourier-transform as well as Fourier-series terms. Their Fourier transforms consist only of equally spaced delta functions of various complex amplitudes. The amplitude of the nth delta

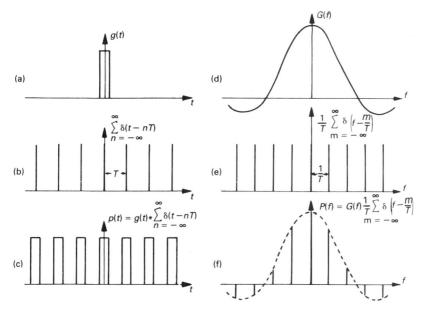

Figure 8.3 An illustration of the FT of a periodic function $p(t)$. (d), (e) and (f) are the FTs of (a), (b) and (c) respectively. (a) shows the motif to be repeated, (b) is a comb defining the periodicity and (c) is the convolution of (a) with (b). with (b). (d) is the FT of the motif, (e) is the comb in f space defining the component frequencies present, and (f) is the product of (d) and (e).

function (i.e. that at frequency n/T) can be shown to be that of the nth harmonic in the exponential Fourier-series representation (7.13).

8.4 Modulation

In the transmission of information by radio waves the original signal is not sent out directly, but is first modified by a process called *modulation*. At the receiving end the incoming wave is *demodulated*, producing a replica of the original signal. There are various types of modulation, the three most common of which, namely *amplitude* modulation, *phase* modulation and *frequency* modulation, will be described briefly in this section.

8.4.1 Amplitude modulation

This is a common technique used in the radio broadcasting of audio signals. It would in principle be possible to convert an audio signal into an electromagnetic wave with the use of an appropriate transducer, but since we require an aerial of length of the order of λ to deliver a significant amount of power in propagating the wave, an impracticably long aerial would be needed at these frequencies. For example, to radiate a signal of 1000 Hz efficiently as a radio signal, we would need an aerial of length of the order of

$$\frac{\text{velocity of light}}{\text{frequency}} \approx \frac{3 \times 10^8}{1000} = 3 \times 10^5 \text{ m} - \text{about two hundred miles!}$$

It is evident, therefore, that it is prohibitively difficult to transmit audio signals as electromagnetic waves; the technique of *amplitude modulation* is one in which the audio signal is processed in such a way as to change its frequency spectrum so that only much higher frequencies are in the transmitted signal. An acoustic signal consists of a power spectrum (of positive and negative frequencies) centred on zero frequency. We shall see in some detail below that the technique of *amplitude modulation* enables one to shift this spectrum bodily so that it becomes centred at any desired higher frequency; the signal corresponding to the modified spectrum can then be transmitted efficiently.

When no information is being transmitted, the transmitter gives out a pure sinusoidal electromagnetic wave known as the *carrier*. Different broadcasting stations operate at different *carrier frequencies*, each of which is characteristic of a particular station. When the input stage of the receiving set is adjusted to be at resonance with the carrier frequency, the set is said to be 'tuned' to that station. The carrier frequencies are (by audio standards) extremely high, typically in the range 0.1–100 MHz. Now suppose it is desired to send an audio signal $y(t)$ (the spectrum of which lies between about 30 Hz and 15 kHz) by radio. It is first necessary that the magnitude of the signal $|y(t)|$ never exceeds a certain value. Let us suppose, for simplicity, that this value is unity. So $|y(t)| \leqslant 1$. The initial part of the process of amplitude modulation is to add a d.c. component to the signal, so that it is always positive. This can, of course, be achieved by adding such a component of unit amplitude. The signal has therefore now become $1 + y(t)$. The next part of the process is to feed this latter signal into a circuit which gives, as output, the product of the signal with the function describing the carrier wave, namely

$$[1 + y(t)] \cos 2\pi f_\text{c} t, \tag{8.17}$$

where f_c is the frequency of the carrier. Expression (8.17) is the form of the transmitted wave (see Fig. 8.4).

To understand why this should be a desirable modification of the original signal $y(t)$ before transmitting it, it is necessary to know the spectrum of expression (8.17). Before we find this, let us take a simple example. Suppose we wish to transmit a pure tone of frequency f_p; that is, suppose

$$y(t) = \cos 2\pi f_\text{p} t.$$

Then (8.17) becomes

$$(1 + \cos 2\pi f_\text{p} t) \cos 2\pi f_\text{c} t, \tag{8.18}$$

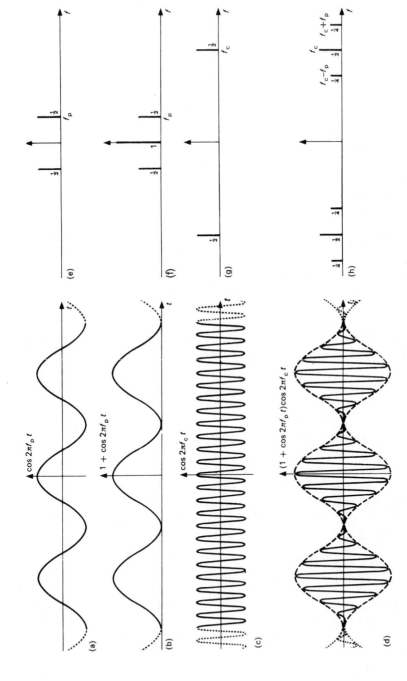

Figure 8.4 The functions on the left-hand side are those used in the explanation in the text of amplitude modulation; their Fourier transforms are shown on the right-hand side.

which, by a little rearrangement involving familiar trigonometric identities, further becomes

$$\tfrac{1}{2} \cos 2\pi(f_c - f_p)t + \cos 2\pi f_c t + \tfrac{1}{2} \cos 2\pi(f_c + f_p)t.$$

The spectrum of this, obtained by Fourier transformation, is

$$\tfrac{1}{4}\delta(f - f_c + f_p) + \tfrac{1}{4}\delta(f + f_c - f_p)$$

$$+ \tfrac{1}{2}\delta(f - f_c) + \tfrac{1}{2}\delta(f + f_c) + \tfrac{1}{4}\delta(f - f_c - f_p) + \tfrac{1}{4}\delta(f + f_c + f_p),$$

by equation (7.22), and is illustrated in Fig. 8.4 together with the spectra of $\cos 2\pi f_p t$, $1 + \cos 2\pi f_p t$ and $\cos 2\pi f_c t$.

The above analysis, together with Fig. 8.4, illustrates many of the essential properties of the amplitude-modulated carrier. We see immediately from the spectrum (h) of the transmitted signal (d) (expression (8.18)) that there is no amplitude at audio frequencies, although the original signal modulating the wave is of audio frequency. This reveals the importance of modulation techniques; frequencies are *changed* from their original values to any value we care to specify, the new frequencies being determined by that of the carrier.

Consider the detailed structure of the spectrum in Fig. 8.4h. The amplitude appears to be concentrated symmetrically about the carrier frequency. This in fact is generally true, as we shall shortly show. The components represented by the delta functions at frequencies $f_c - f_p$ and $f_c + f_p$ are known as *sidebands*. The process of demodulating at the receiving end is essentially one of recovering the original spectrum from the information contained around the region of the carrier frequency. We will, however, not go into details here.

Let us now consider the Fourier transform of the function

$$[1 + y(t)] \cos 2\pi f_c t, \tag{8.17}$$

where $y(t)$ is a general function of time (Fig. 8.5d). We note that the function to be transformed is a product of two functions $1 + y(t)$ and $\cos 2\pi f_c t$. So, by the convolution theorem discussed in section 8.3, the FT of the function must be the convolution of the FT of $1 + y(t)$ and that of $\cos 2\pi f_c t$. If we let the FT of $y(t)$ be $Y(f)$, then the FT of $1 + y(t)$ is $\delta(f) + Y(f)$.

The FT of $\cos 2\pi f_c t$ is $\tfrac{1}{2}[\delta(f - f_c) + \delta(f + f_c)]$ (7.22). So the FT of $[1 + y(t)] \cos 2\pi f_c t$ is

$$[\delta(f) + Y(f)] * \tfrac{1}{2}[\delta(f - f_c) + \delta(f + f_c)]$$

$$= \tfrac{1}{2}[\delta(f - f_c) + Y(f - f_c) + \delta(f + f_c) + Y(f + f_c)] \tag{8.19}$$

by (8.11). The relevant functions and their transforms are shown in Fig. 8.5.

From the nature of the Fourier transform, Fig. 8.5h, we see that the bandwidth required to accommodate all the information in the original $y(t)$, when that signal is propagated as an amplitude-modulated wave, is twice the extent of the original spectrum. One of the advantages of the technique is that, by choosing appropriate carrier frequencies, the information can be propagated in whatever frequency range

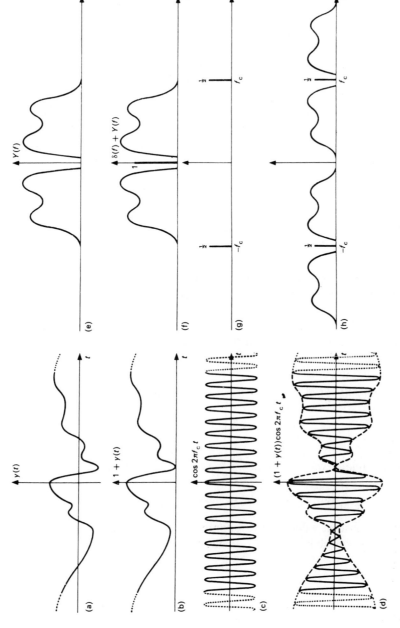

Figure 8.5 The functions on the right-hand side are the Fourier transforms of those on the left. The transmitted signal (d) is the product of (b) and (c); the FT, (h), of the transmitted signal is the convolution of (f) and (g).

we choose. No interference with other information being transmitted on the same, or a neighbouring, transmitter can occur if the carrier frequencies for the two signals are sufficiently far apart in frequency.

Furthermore, it is possible to accommodate twice as many signals in a given frequency range than the above discussion would indicate by suppressing the sidebands of frequencies lower than (or higher than) that of the carrier, the information from the remaining sideband still being extractable. This is because the amplitude spectrum is symmetrical about the carrier frequency; there is therefore no essential information about the signal in the lower sideband that is not already contained in the upper sideband. This *single-sideband* transmission occupies only half the bandwidth taken by the double-sideband modulation, and therefore we can put other information in the suppressed sideband region provided, of course, that the receiving set is constructed accordingly.

8.4.2 Phase modulation and frequency modulation

These are both examples of what is known as *angle modulation*. The signal $y(t)$, instead of modifying the amplitude of the carrier wave, as in the previous case, modifies the angle which is the argument of the cosine function representing the carrier. Without any signal, the carrier can be represented as $\cos 2\pi f_c t$; this is modified, when a signal $y(t)$ phase modulates the carrier, to $\cos[2\pi f_c t + y(t)]$. So we see that the amplitude of the carrier remains constant regardless of the presence or absence of a signal, and the information in the signal is contained in the deviation of the argument of the cosine from what it would be if $y(t)$ were zero. In a very similar way the representation of a carrier wave *frequency modulated* by a signal $y(t)$ is

$$\cos\left[2\pi f_c t + \int_{-\infty}^{t} y(u)\, \mathrm{d}u \right] \tag{8.20}$$

where u is a dummy variable in units of time. The reason for calling (8.20) a frequency-modulated wave is as follows. We can define the *instantaneous frequency* f_i as

$$\frac{1}{2\pi} \frac{\mathrm{d}}{\mathrm{d}t} \text{ (argument of the cosine).}$$

This yields a constant value f_c for an unmodulated cosine wave $\cos 2\pi f_c t$, since the argument of the cosine here is $2\pi f_c t$, giving f_i as

$$f_i = \frac{1}{2\pi} \frac{\mathrm{d}}{\mathrm{d}t} (2\pi f_c t) = f_c.$$

In a similar way, we find the instantaneous frequency for the expression (8.20) is

$$f_i = \frac{1}{2\pi} \frac{\mathrm{d}}{\mathrm{d}t} \left[2\pi f_c t + \int_{-\infty}^{t} y(u)\, \mathrm{d}u \right]$$

$$= f_c + \frac{1}{2\pi} y(t).$$

Thus

$$f_i - f_c = \frac{1}{2\pi} \, y(t),$$

and we see that the instantaneous deviation in frequency from that of the carrier is directly proportional to the magnitude of the signal at that time. Hence the wave (8.20) is described as a frequency-modulated wave.

We do not propose to go into the question of the spectra of these waves; these are much more complicated than for amplitude-modulated waves and we refer the reader who may be interested in the matter to relevant texts.

Problems

1. The derivative theorem (section 8.1.4) was proved by considering a suitable modification of the inverse Fourier transform (7.14). Can you derive the shift theorem in a similar fashion?

2. The relationship between $\delta(t)$ and $\delta(at)$ was treated in problem 8 in Chapter 7. Using the similarity theorem, prove the result that $\delta(at) = \delta(t)/|a|$ (a being a non-zero constant).

3. The function of time t which has a value of unity in the region $-\frac{1}{2} \leqslant t \leqslant +\frac{1}{2}$ and of zero elsewhere is often known as rect(t) because its graph against t looks like a rectangle. Show that its Fourier transform is sinc(f). [It will be recalled from Chapter 7 that sinc(x) is a shorthand notation for $\sin(\pi x)/(\pi x)$.] At what frequencies is there zero amplitude in sinc(f)? Deduce, either from first principles or from the appropriate theorem, that the FT of rect(at) is $(1/a)$ sinc(f/a), where a is a positive constant. You will find rect(t), rect($2t$) and their transforms sketched in Fig. 8.1.

4. Show that $\int_{-\infty}^{\infty} g(t) \, dt$ is equal to the value of the FT of $g(t)$ at zero frequency, i.e. $G(0)$. Show also that

$$\int_{-\infty}^{\infty} G(f) \, df = g(0).$$

The integral $\int_{-\infty}^{\infty} \text{sinc}(x) \, dx$ is rather difficult to evaluate directly. However, using the results of problem 3, deduce that its value is unity.

5. It is usually true that the convolution of two functions is broader than either function; for this reason, convolution is sometimes known as *smearing*. This breadth increase is, however, no more than a rule of thumb; indeed, there exist pathological cases in which the convolution is narrower than either function. The following problem deals with the curious special case in which the convolution is actually equal to whichever of the two functions is the broader.

Consider the convolution in frequency space

$$[(1/a) \, \text{sinc}(f/a)] * [(1/b) \, \text{sinc}(f/b)],$$

where *a* and *b* are positive constants. Without putting it into the integral form provided by the definition (8.8) of convolution (for you will get yourself into a fearful mess if you do), but instead referring to problem 3, show that the convolution is equal to $(1/a)$ sinc(f/a) if $b \geqslant a$ and $(1/b)$ sinc(f/b) if $a \geqslant b$.

6. A businessman needs to transmit a large quantity of verbal information by long-distance telephone. To save money, he records this information on a tape recorder, and plays it back at ten times the speed over the telephone. The recipient records this, and plays it back at one-tenth of the speed. If the bandwidth for intelligible speech is f_1 to f_2, what bandwidth must be possessed by the recorders and telephone line?

Telephone lines are inherently 'noisy'. Suggest why the process described above would produce a 'noisier' signal than if the telephone message had been delivered normally.

7. The FT of the function $\exp(-\pi t^2)$ is $\exp(-\pi f^2)$ (a result very easy to remember!). Use the similarity theorem to deduce the FT of $\exp(-\alpha t^2)$ where α is a positive constant. Use the derivative theorem to deduce that of $t \exp(-\alpha t^2)$ and $t^2 \exp(-\alpha t^2)$.

8. Deduce the current $i(t)$ in the *RLC* circuit of Example 8.1, for the case $v(t) = v_0 \cos(2\pi f_0 t)$, $(-\infty < t < +\infty)$.

9. Using the FT of $\exp(-\alpha t^2)$ from problem 7 and the convolution theorem, show that

$$\exp(-x^2/a^2) * \exp(-x^2/b^2) = ab[\pi/(a^2 + b^2)]^{1/2} \exp[-x^2/(a^2 + b^2)].$$

[The function $\exp(-\alpha t^2)$ is known as the *Gaussian function* and a normalised version of it is the probability density function (p.d.f.) of a commonly occurring distribution called the *normal distribution*. The above convolution describes the compounding of the p.d.f.s of two independent normal distributions whose random variables are added. You see that the result of the convolution in this case is also a normal distribution.]

10. The sound-track of a film is a strip situated near the edge of the film. The sound information $x(t)$ is encoded on the strip as exposure of variable density. The film goes at a uniform velocity past an illuminated narrow slit, the illumination being sensed by a photoelectric cell and converted into an electrical signal $y(t)$. This signal is then fed into an amplifier and loudspeaker. If the short time taken for the film to pass from one side of the slit to the other is T, then the slit is behaving as a 'window', $h(t) = \text{rect}(t/T)$. It can be shown that the output $y(t)$ is the convolution of the input $x(t)$ with the slit function $h(t)$, i.e.

$$y(t) = x(t) * h(t).$$

[This equation can be applied, with varying degrees of approximation, to many situations. The input $x(t)$ is degraded by convolution (smearing) with a function $h(t)$ which is entirely to do with the instrument (in this case the projector) through which it is passing. The convolution operation has the effect of 'factorising' the input itself from the effect that the instrument has upon the input.]

The viewer of the film would, of course, like to hear the signal $x(t)$, but must perforce put up with the degraded version $y(t)$. Using the convolution theorem, deduce the FT, $Y(f)$, of the output in terms of the FTs, $X(f)$ and $H(f)$, of the input and instrument response function respectively. Using the result of problem 3, deduce $H(f)$. What frequency components in the signal will never get through the system regardless of how prominent they are in the spectrum $X(f)$? If it is desired to amplify up to frequencies of 15 000 Hz and the speed of the film is 40 cm s^{-1}, then estimate the maximum width that the slit may have. Can you think of any modification to the slit which would lessen the degradation of the signal upon convolution?

11. A certain 'black box' accepts an input signal $x(t)$ and squares it to produce an output $y(t) = x^2(t)$. Use the convolution theorem to determine how the spectrum $Y(f)$ is related to the spectrum $X(f)$ of the input. What frequencies are present in the output if the signal is (i) $\cos(2\pi f_0 t)$ and (ii) $\cos(2\pi f_1 t) + \cos(2\pi f_2 t)$?

9 Sound

9.1 Introduction

This chapter is concerned with some features of sound waves. The word 'sound' has two distinct meanings. The first meaning is 'longitudinal waves in matter' – that is, the actual disturbance and its propagation in the medium; the second meaning is 'those longitudinal waves in a fluid which evoke a percept in the auditory system'. Now, as we shall discuss later in this chapter, the auditory system (i.e. the combination of ears and brain) can perceive only a limited range of frequencies. The range varies from person to person and is also dependent upon the age of a person, but is usually from about 16 to 16 000 Hz. It is therefore evident that there can exist 'sound' in the sense of the first meaning which is certainly not sound according to the second. For this reason, physicists and others have, in recent years, referred to sounds of very low, inaudible frequencies as *infrasonic* and sounds of very high, inaudible frequencies as *ultrasonic*. But audibility is not determined solely by frequency, because, even though a sound has a spectrum within the audible frequency range, it does not necessarily mean that it can be heard. In fact for amplitudes below a certain 'threshold' the sound is too faint to be heard, i.e. the wave is not sufficiently energetic to evoke a percept.

The first observation to be made about sound in air (in its first meaning) is that it is a longitudinal wave motion that must be described by a solution of the three-dimensional equivalent of the one-dimensional wave equation

$$\frac{\partial^2 z}{\partial t^2} = c^2 \frac{\partial^2 z}{\partial x^2} \tag{9.1}$$

discussed in Chapter 6. Sound waves in air have to be longitudinal because the air is unable to sustain a torsional or shear force.

In this equation, z is the displacement at a position x and at a time t, and c, the velocity of the sound waves, is $\sqrt{(\gamma P/\rho)}$, where γ is the ratio of principal specific heats for air, P is the ambient pressure and ρ is the density. The velocity of sound is about 340 m s^{-1} (760 m.p.h.); that is, a sound wave travels roughly one-third of a kilometre in one second. Since this velocity is so much less than that of light, distant events producing simultaneous optical and acoustical disturbances (such as in an electrical storm) are heard later than they are seen.

9.2 Sound waves in a pipe

Sound waves within a pipe of fixed length must, of course, be described by solutions of (9.1); however, the presence of the pipe imposes boundary conditions on the problem which restrict the possible solutions, as was described in Chapter 5. Before embarking on an analysis of sound waves in pipes we make two assumptions, of which the first is that the length L of the pipe is much larger than its radius. This assumption enables us to consider the pipe as essentially a one-dimensional entity, in which the air can vibrate only along its length. We therefore are able to use the one-dimensional wave equation (9.1). We also assume that the pipe is of uniform cross-section.

We shall consider in detail the pipe closed at one end and open at the other. The x-axis is taken along the pipe. There are two boundary conditions that must be applied. The first is very simple; at no time can there be any displacement at the closed end of the pipe, at $x = 0$, because this is considered to be rigidly fixed. Our first boundary condition therefore is

$$[z]_{x=0} = 0 \quad \text{for all values of } t. \tag{9.2}$$

The second boundary condition is concerned with the magnitude of vibrations at the open end of the pipe, and is by no means as simple as the first. It will be recalled from Chapter 6 that the displacement variations in a sound wave are accompanied by pressure variations. Indeed the sound wave can be specified either by the pressure or the displacement variations, as desired. As was shown in section 6.6, the acoustic pressure p and the displacement z are related by

$$p = -K \frac{\partial z}{\partial x}. \tag{6.24}$$

Now, because of the presence of the rigid wall of the pipe, there is no opportunity for any excess pressure caused by the displacement to dissipate in a direction perpendicular to the length of the pipe. However, beyond the open end of the pipe there is such an opportunity and, consequently, the acoustic pressure falls away very quickly. So to a first approximation we can say that there is no acoustic pressure outside the pipe. This, as we shall discuss later on, is an assumption we can make only with reservations, but it is not unreasonable. So our second boundary condition is that

$$[p]_{x=L} = 0 \quad \textit{for all values of } t,$$

which, because of (6.24) can be re-expressed as

$$\left[\frac{\partial z}{\partial x} \right]_{x=L} = 0 \quad \text{for all values of } t. \tag{9.3}$$

The wave equation (9.1), together with the boundary conditions (9.2) and (9.3), enable us to deduce a complete description of the wave motion within the pipe. Whatever the most general motion is, we can say from our experience in Chapter 2 that it must be some linear combination of the normal modes. Since the air in the pipe can be regarded as a continuous medium, there must be an infinite number of normal modes of vibration. Let us therefore find a typical normal mode solution. It will be recalled that a normal mode solution is one for which the air vibration throughout the whole pipe takes place at a single frequency, say f_n (the suffix n indicates the nth normal mode). The amplitude of vibration will vary along the length of the pipe but will, of course, be constant with time; we will describe its variation by the function $\phi_n(x)$. So a typical normal mode solution is

$$z_n(x, t) = \phi_n(x) \cos(2\pi f_n t + \alpha_n), \tag{9.4}$$

where α_n is a constant phase angle which belongs to the nth normal mode. Partially differentiating (9.4) we obtain

$$\frac{\partial^2 z_n}{\partial t^2} = -4\pi^2 f_n^2 \phi_n(x) \cos(2\pi f_n t + \alpha_n)$$

and

$$\frac{\partial^2 z_n}{\partial x^2} = \cos(2\pi f_n t + \alpha_n) \frac{d^2 \phi_n(x)}{dx^2}.$$

Substituting these derivatives into the original wave equation (9.1) we obtain

$$-4\pi^2 f_n^2 \phi_n(x) \cos(2\pi f_n t + \alpha_n) = c^2 \cos(2\pi f_n t + \alpha_n) \frac{d^2 \phi_n(x)}{dx^2},$$

which, after some cancellation and rearrangement, becomes

$$\frac{d^2 \phi_n(x)}{dx^2} = -\frac{4\pi^2 f_n^2}{c^2} \phi_n(x). \tag{9.5}$$

Equation (9.5) is an ordinary differential equation in the one independent variable x, and we see immediately that it is identical in form to the simple-harmonic motion equation, except that the latter has t as its independent variable. The general solution of this has the form given by (2.2), namely

$$\phi_n(x) = a_n \sin\left[\left(\frac{4\pi^2 f_n^2}{c^2}\right)^{1/2} x + \varepsilon_n\right]$$

$$= a_n \sin\left[\left(\frac{2\pi f_n}{c}\right) x + \varepsilon_n\right],$$

where a_n and ε_n are arbitrary constants which are, in general, different for different values of n. Thus the amplitude variation $\phi_n(x)$ repeats periodically at intervals of

$$\lambda_n = \frac{2\pi}{2\pi f_n/c} = \frac{c}{f_n}.$$

So

$$c = f_n \lambda_n \tag{9.6}$$

as expected.

The normal mode solution is, therefore, one of sinusoidal form (although we have not yet determined its phase) and we have shown that the wavelength λ_n is related to the frequency f_n by the simple, familiar relationship (9.6). Since the amplitude a_n is completely arbitrary, any amplitude of vibration is possible. In other words the wave equation determines only the *form* of the wave, not its *scale*.

The typical normal mode solution (9.4) can now be written as

$$z_n(x, t) = a_n \sin\left[\left(\frac{2\pi f_n}{c}\right)x + \varepsilon_n\right] \cos(2\pi f_n t + \alpha_n). \tag{9.7}$$

This will also be recognised as being of the form of a standing wave (compare (9.7) with (5.6)) as expected, since there are two waves moving in opposite directions.

Let us now apply the boundary condition (9.2), namely

$$[z_n]_{x=0} = 0 \quad \text{for all } t;$$

this gives

$$z_n(0, t) = a_n \sin \varepsilon_n \cos(2\pi f_n t + \alpha_n) = 0.$$

For this to be zero for all values of t, $a_n \sin \varepsilon_n$ must be zero. Since $a_n = 0$ represents the trivial case of no vibration at all, we are forced to conclude that $\sin \varepsilon_n$ is zero. This implies that $\varepsilon_n = m\pi$, where m is an integer. The only physical distinct values of m are 0 and 1, since the angles $2\pi, 4\pi, \ldots$ are identical to 0, and $3\pi, 5\pi, \ldots$ to π. The case $\varepsilon_n = \pi$ merely changes the sign of the amplitude a_n, so, without any loss of generality, we may take ε_n as zero. Thus (9.7) may be rewritten as

$$z_n(x, t) = a_n \sin\left(\frac{2\pi f_n}{c}\right)x \cos(2\pi f_n t + \alpha_n). \tag{9.8}$$

To apply the second boundary condition (9.3),

$$\left[\frac{\partial z_n}{\partial x}\right]_{x=L} = 0 \tag{9.3}$$

we must partially differentiate (9.8) with respect to x This gives us

$$\frac{\partial z_n(x, t)}{\partial x} = a_n 2\pi \frac{f_n}{c} \cos\left(2\pi \frac{f_n}{c} x\right) \cos(2\pi f_n t + \alpha_n),$$

which, for $x = L$, becomes

$$\left[\frac{\partial z_n(x, t)}{\partial x}\right]_{x=L} = a_n 2\pi \frac{f_n}{c} \cos\left(2\pi \frac{f_n}{c} L\right) \cos(2\pi f_n t + \alpha_n).$$

But this must be zero for all values of t, by equation (9.3), and so

$$\cos 2\pi \frac{f_n}{c} L = 0.$$

This implies that

$$2\pi \frac{f_n}{c} L = (n + \tfrac{1}{2})\pi,$$

where n is an integer, i.e.

$$\frac{f_n}{c} = \frac{(n + \tfrac{1}{2})}{2L}$$

$$= \frac{(2n + 1)}{4L}. \tag{9.9}$$

Thus the functions $\phi_n(x)$ are

$$\phi_n(x) = a_n \sin \frac{2\pi(2n + 1)x}{4L},$$

which are sine functions of wavelengths

$$\lambda_0 = 4L, \ \lambda_1 = \tfrac{4}{3}L, \ \lambda_2 = \tfrac{4}{5}L, \ \lambda_3 = \tfrac{4}{7}L, \text{ etc.}$$

The first four of these functions are sketched in Fig. 9.1, from which it will be seen that an antinode of displacement exists at $x = L$ for all the functions ϕ_n. This is a direct consequence of the pressure node at the open end of the pipe which was the second of our boundary conditions. The displacement nodes at $x = 0$ for all the ϕ_n are, of course, a result of the boundary condition of no displacement at the closed end. The values of the different f_n as seen from (9.9) to be

$$f_n = \frac{(2n + 1)}{4L} c;$$

substituting $n = 0$ in this equation we obtain $f_0 = c/4L$. Similarly $f_1 = 3c/4L$, $f_2 = 5c/4L$, etc.

The relationship between the natural frequencies is therefore

$$f_1 = 3f_0, \ f_2 = 5f_0, \ f_3 = 7f_0, \text{ and so on.}$$

These frequencies are thus seen to be very simply related to each other; in fact if we had analysed the wave motion in a pipe open at both ends, we would have found an even simpler relationship, namely that the frequencies would have been related in the ratios of the natural numbers 1, 2, 3, 4, ...; that is, $f_2 = 2f_1$, $f_3 = 3f_1$, and so on. The reason why we did not analyse this case was that it did not give an opportunity to apply the two different kinds of boundary condition provided by the pipe closed at one end and open at the other.

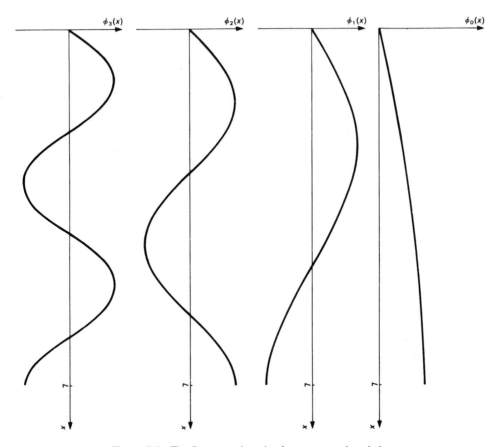

Figure 9.1 The first normal modes for an open–closed pipe.

We are now in a position to see why the pipe is such an important constituent part of some musical instruments. Taking the open–open pipe as our example, we have already seen that the normal mode frequencies are f_1, $2f_1$, $3f_1$, $4f_1$, etc. This is the *harmonic* series with f_1, the *first harmonic* (or *fundamental*), f_2 the *second harmonic* and so on. If the first eight of these could be successively sounded in ascending order of harmonics, and if the fundamental happened (for example) to be C below the bass stave, the musical effect would be as follows:

With the exception of the seventh harmonic, which is an excruciatingly flat B♭ (for which reason we have put the note in brackets) the first eight harmonics are all notes to be found on the western scale; in fact, the musical reader will immediately recognise them as forming the common chord of C major.

Since the general state of motion of the air in any pipe must be the sum of all the normal mode solutions in varying strengths, depending on the method of exciting the motion, a given pipe can give out only frequencies in the harmonic sequence. These frequencies (for reasonably small harmonic number, at any rate) sound harmonious when heard in conjunction with each other. The musical nature of the sound from a pipe depends entirely on the fact that the natural frequencies for this acoustic system happen to be harmonically related to each other. The reason why a drum does not sound as musical as a pipe is that the natural frequencies of vibration of the drum head are related to each other in a very complicated way (in fact it may be shown that $f_2 \cong 1.5933f_1$, $f_3 \cong 2.1355f_1$, $f_4 \cong 2.2954f_1$, and so on) and the effect of these when heard simultaneously is nowhere near as harmonious as that from a pipe.

A little more should be said about the boundary condition (9.3) for the open end of a pipe. Strictly speaking, the problem is much more complex than was apparent from our analysis. There is some acoustic pressure just outside the open end, and our justification for assuming there was none was that the acoustic pressure outside was much less than that inside. More accurate (and complicated) analysis shows that the approximation is such that the pipe appears to be operating with an effective length slightly larger than its physical length. The extra length to make the frequencies observed fit the theory has been shown from experimentation and theoretical analysis to be about $0.6R$ where R is the radius of the pipe. This is known as the *end-correction*. Further, the end-correction has been shown to depend very slightly on the wavelength, with the consequence that the harmonics are progressively more out of tune with an integral multiple of the fundamental as the harmonic number increases. But this is not, in most cases, a large enough deviation to worry the listener.

Another way in which the theory is approximate is that we have tacitly assumed that no acoustical energy is lost from the pipe, so that, once we have excited the air in the pipe, the vibrations will go on for an infinite time, even if we terminate the excitation. That this is not so is a matter of common observation; organ pipes, for example, cease to speak as soon as the organist takes his or her fingers (or feet) off the keys. In other words, the resonances within the pipe are, in fact, not the perfect ones implied in the foregoing analysis. In practice, acoustical energy is being lost from the pipe by radiation to the surrroundings. It is this loss which enables one to hear the instrument at a distance. A compromise is occurring here. On the one hand, if the resonances were very good (i.e. perfect boundary conditions), as would occur in a closed–closed pipe, the sound would be musically very pure, but none of it would propagate from the interior of the pipe and no sound would be heard from a distance at all! On the other hand, if the resonances were very imperfect (i.e. of low Q), the radiation from the pipe would be more efficient, and the hearer would perceive plenty of unmusical sound! In a musical instrument, it is desirable to be as near the former case as practicable in order for the sound to have a distinct pitch. In a loudspeaker, on the other hand, resonances are undesirable because some frequencies would propagate more efficiently than others; also, generally efficient propagation is desired. So the designer of a loudspeaker needs to approach as nearly as possible to the latter case. Loudspeakers are therefore not cylindrical, as are most pipes in musical instruments, but may, for example, be conical or exponential in shape.

A final word should perhaps be said about the open–closed pipe. These are found, as well as open–open pipes, in organs, and they produce a tone that is significantly different from that given by the open–open pipe, since only the odd harmonics are present. They are also more economical to build than open–open pipes by the following reasoning. We found the fundamental frequency of the open–closed pipe to be $c/4L$. A similar analysis of an open–open pipe would reveal a fundamental of $c/2L$. Therefore a given fundamental is produced by an open–closed pipe of half the length of its open–open counterpart. So an open–closed pipe of what an organ builder or organist would describe as '8-foot pitch' is actually only 4 ft (1.2 m) in physical length. Therefore, the open–closed pipe can be constructed from about half the material needed for the open–open pipe of the same fundamental. This can become a considerable economic consideration in the case of pipes of 32-foot pitch. Organ builders cannot dispense entirely with open–open pipes however, on account of the fuller tone provided by the presence of all the harmonics.

We state finally that the clarinet's distinctive tone is connected with the fact that, to a good approximation, it behaves as a closed pipe.

9.3 Waves on strings

We have said a great deal about transverse waves on strings as easily visualised examples of wave motion; our reason for discussing them still further is that strings, as well as pipes, form the bases of many musical instruments such as the guitar, pianoforte, harp, members of the violin family and so forth. They are thus very relevant to a discussion of sound.

Let us consider a string of mass per unit length μ, length L, under a tension T, fixed at both ends as shown in Fig. 9.2. The wave equation (9.1) is valid for the stretched string, as we saw in Chapter 6, the velocity c, of transverse waves, in this case being $\sqrt{(T/\mu)}$. Our analysis will proceed much as that for waves in a pipe. First we need the boundary conditions; clearly, these are that the displacement y at each end of the string must be zero at all times. That is,

$$y(0, t) = 0 \tag{9.10}$$

and

$$y(L, t) = 0. \tag{9.11}$$

We take a typical normal mode solution of (9.1) to be

$$y(x, t) = \phi_n(x) \cos(2\pi f_n t + \alpha_n) \tag{9.12}$$

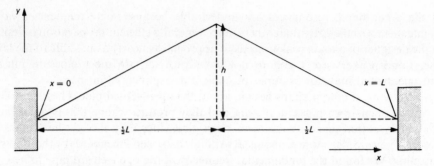

Figure 9.2   An example of initial conditions for a stretched string.

and, by an argument precisely the same as that in the last section, we can obtain that $\phi_n(x)$ is the solution to the ordinary differential equation

$$\frac{d^2\phi_n(x)}{dx^2} = -\frac{4\pi^2 f_n^2}{c^2}\,\phi_n(x).$$

As before, therefore, the most general solution is

$$\phi_n(x) = a_n \sin\left(\frac{2\pi f_n}{c}x + \varepsilon_n\right),$$ (9.13)

and substitution of the boundary condition (9.10) yields the result $\varepsilon_n = 0$. But the second boundary condition (9.11) differs from that in the previous section, and its implications can be seen by equating $\phi_n(L)$ in (9.13) to zero. This gives

$$a_n \sin \frac{2\pi f_n}{c} L = 0,$$

i.e.

$$\frac{2\pi f_n}{c} L = 0,\ \pi,\ 2\pi,\ \ldots,\ n\pi,\ \ldots :$$

Thus the natural frequencies f_n are $nc/2L$. Since $c = \sqrt{(T/\mu)}$, the natural frequencies of the string are finally

$$f_n = \frac{n}{2L}\sqrt{\frac{T}{\mu}}.$$ (9.14)

Equation (9.14) is an expression of *Mersenne's law*; we see immediately from it that the natural frequencies are a harmonic series, since

$$f_1 = \frac{1}{2L}\sqrt{\frac{T}{\mu}}.$$ (9.15)

and therefore $f_n = nf_1$.

The general motion of the string is a linear combination of all the normal modes, and therefore a string of given mass per unit length, under a given tension and of fixed length, can give only a combination of all its harmonics and no other frequency. This property, as we saw in the case of the pipe in the last section, is that which makes the string so suitable as the primary vibrator in a musical instrument.

Mersenne's law determines to a considerable extent the design, and in some cases the shape, of stringed musical instruments. We see from (9.15) that the fundamental frequency is determined exclusively by T, μ and L. Equation (9.15) determines, for example, the combination of the basic shape of the grand pianoforte, and the gradation in mass per unit length from string to string along the compass of the instrument. Suppose the lengths of all the strings in the pianoforte were the same, and further suppose all the strings were from one reel of wire. Then μ and L would be constants for the instrument, and the only way in which we could produce all the different notes over the instrument's seven-octave range would be by having the strings at different tensions.

Now, each octave rise in pitch means a doubling of the frequency; this means that the highest note on the instrument is of frequency 2^7 times that of the lowest note seven octaves lower. If the frequencies of the top and bottom note are denoted by f_t and f_b respectively, with corresponding string tensions T_t and T_b, then (9.15) tells us that, for constant L and μ,

$$\frac{f_t}{f_b} = \sqrt{\frac{T_t}{T_b}}.$$

Thus

$$\frac{T_t}{T_b} = \left(\frac{f_t}{f_b}\right)^2 = (2^7)^2 = 2^{14} = 16\,384.$$

So, whatever other defects this arrangement may have, the frame of the instrument would have an absurdly uneven distribution of tension. Clearly, some other way has to be found to give us the seven-octave range. In practice, all the possible variables, T, μ and L, are used to this end, but the gradations of μ and L are the most obvious. The strings are shorter for higher frequencies than for lower, giving the grand pianoforte its harp-like shape (the harp's shape is, of course, determined by the same reasoning). In addition the strings of the lower notes are thicker than those of the higher notes, having the effect of reducing the enormous length the instrument would otherwise have.

The increased mass per unit length for the lower strings is achieved in a rather interesting way. The strings are not simple thick strings; they are comparatively thin strings which are overspun – that is, a second wire is coiled tightly round the original wire over its whole length. The effect of this is to make the string flexible; a simple thick string would have somewhat rod-like characteristics so that transverse waves along it would, as was pointed out briefly in Chapter 1, be dispersive, causing the natural frequencies of the string not to be harmonically related to each other. In fact, even with overspun strings and simple strings for the higher notes, the natural frequencies are not harmonic because of stiffness (i.e. non-flexibility), but this defect is not sufficiently large to be noticed directly by the ear. Nevertheless it is a fact that a pianoforte, tuned so that the fundamental frequencies of all the notes are in tune, sounds badly out of tune, particularly at the higher end, and it may well be that the ear is taking as its criterion of pitch the upper harmonics of notes in the middle and lower ranges, rather than the fundamentals.

Mersenne's law (9.14) also determines some of the characteristics of instruments of the violin family. In these instruments, as in the case of the guitar and some other instruments, the different notes of the scale are obtained not by having a different string for each note, but by having a limited number of strings (only four in the case of the violin) whose effective lengths L are varied by 'stopping' the strings with the fingers of the left hand. The range of the instrument would be somewhat limited if only one string were used, and one of the purposes of the other three strings is to increase it. (Another reason for having more than one string is that several notes can be sounded simultaneously; this facility is extensively used in guitar music.) Furthermore, since the basic design is such that the four strings be of the same length, either the tension or the mass per unit length must be different for each string. Since it is undesirable that the tension be significantly different, causing the bridge of the instrument to be unevenly stressed, it is the mass per unit length that is different in practice. It is therefore necessary to buy a G-, D-, A- and E-string for the instrument, all of which are of different masses per unit length. The G-string, which produces the lowest fundamental, is usually overspun with fine silver wire to increase μ, for the same reason as for the lower notes on the pianoforte. The D- and A-strings are usually made of gut (of different thicknesses) and the E-string, which has the highest fundamental, is usually made of fairly fine steel wire which combines the necessity for a small value of μ with that of mechanical strength.

We see, therefore, that Mersenne's law is basic to many aspects of the design of stringed instruments in that it determines the conditions which must be satisfied to obtain a desired fundamental frequency.

We will return to some other aspects of musical instruments later in this chapter, but let us for the present discuss some further properties of vibrating strings. We know that the general motion of a string must be some linear combination of its normal modes, and that, given the initial conditions, the strengths of the various harmonics can be determined. The typical normal mode solution is (from (9.12) and (9.13))

$$y_n(x, t) = a_n \sin\left(2\pi \frac{f_n}{c} x\right) \cos(2\pi f_n t + \alpha_n),$$

and since $f_n/c = n/2L$ and $f_n = nf_1$, from Mersenne's law (9.14),

$$y_n(x, t) = a_n \sin \frac{\pi n x}{L} \cos(2\pi n f_1 t + \alpha_n).$$

The general motion of the string is

$$y(x, t) = \sum_{n=1}^{\infty} a_n \sin \frac{\pi n x}{L} \cos(2\pi n f_1 t + \alpha_n). \tag{9.16}$$

We can re-express (9.16) as

$$y(x, t) = \sum_{n=1}^{\infty} a_n \sin \frac{\pi n x}{L} (\cos 2\pi n f_1 t \cos \alpha_n - \sin 2\pi n f_1 t \sin \alpha_n).$$

We can replace $a_n \cos \alpha_n$ by C_n and $-a_n \sin \alpha_n$ by S_n, which gives

$$y(x, t) = \sum_{n=1}^{\infty} \sin \frac{\pi n x}{L} (C_n \cos 2\pi n f_1 t + S_n \sin 2\pi n f_1 t). \tag{9.17}$$

The coefficients C_n and S_n depend on the initial conditions of the string, that is, on the positions and velocities of each point on the string at time $t = 0$. These vary, of course, according to the particular problem. For example, for a pianoforte string, the initial displacements $y(x, 0)$ are zero, but the initial velocities depend on the shape, size and elastic properties of the hammer, and the velocity with which it hits the string. We will, however, not attempt to analyse this difficult problem; instead we shall take the problem of a string constrained into some shape described by the function $y(x, 0)$, and then, at time $t = 0$, released. Our problem is to find the subsequent motion of the string with the aid of (9.17).

The initial velocity of every point on the string is zero; our initial conditions can therefore be summarised by the fact that the initial shape is $y(x, 0)$ and that

$$\left[\frac{\partial y}{\partial t} \right]_{t=0} = 0, \tag{9.18}$$

for all x.

Let us first find the effect of the initial condition (9.18) upon (9.17). If we differentiate this latter equation partially with respect to time we obtain

$$\frac{\partial y}{\partial t} = \sum_{n=1}^{\infty} \sin \frac{\pi n x}{L} (-2\pi n f_1 C_n \sin 2\pi n f_1 t + 2\pi n f_1 S_n \cos 2\pi n f_1 t),$$

which, for time $t = 0$, becomes

$$\left[\frac{\partial y}{\partial t} \right]_{t=0} = \sum_{n=1}^{\infty} 2\pi n f_1 S_n \sin \frac{\pi n x}{L}.$$

Since this must be zero for all value of x, by (9.18), it follows that all the coefficients S_n must be zero. So (9.17) assumes the simpler form

$$y(x, t) = \sum_{n=1}^{\infty} C_n \sin \frac{\pi n x}{L} \cos 2\pi n f_1 t. \tag{9.19}$$

To find the C_n we apply our knowledge of the initial shape $y(x, 0)$ to (9.19). At time $t = 0$, this equation becomes

$$y(x, 0) = \sum_{n=1}^{\infty} C_n \sin \frac{\pi n x}{L}. \tag{9.20}$$

Suppose we wish to find a particular coefficient C_m; following the methods developed in section 7.1 we merely multiply (9.20) by

$$\sin \frac{\pi m x}{L} \, \mathrm{d}x$$

and integrate over the whole length of the string, obtaining

$$\int_0^L y(x, 0) \sin \frac{\pi m x}{L} \, \mathrm{d}x = \int_0^L \sin \frac{\pi m x}{L} \sum_{n=1}^{\infty} C_n \sin \frac{\pi n x}{L} \, \mathrm{d}x.$$

Just as in our discussions on Fourier series, the nth term on the right-hand side is zero unless $n = m$, and the equation therefore reduces to

$$\int_0^L y(x, 0) \sin \frac{\pi m x}{L} \, \mathrm{d}x = C_m \int_0^L \sin^2 \frac{\pi m x}{L} \, \mathrm{d}x.$$

The integral on the right-hand side of the latter equation can easily be shown to be $\frac{1}{2}L$, reducing the equation still further to

$$\int_0^L y(x, 0) \sin \frac{\pi m x}{L} \, \mathrm{d}x = \frac{1}{2} C_m L.$$

Finally, reverting to the nomenclature C_n rather than C_m, we have

$$C_n = \frac{2}{L} \int_0^L y(x, 0) \sin \frac{\pi n x}{L} \, \mathrm{d}x. \tag{9.21}$$

So in the case of zero initial velocity, all we need do to calculate the C_n of (9.19) is to use our knowledge of the initial shape of the string, $y(x, 0)$, together with the integral in (9.21). It may seem surprising that Fourier series enter into this problem, and also that they arise in a spatial rather than a temporal way. This is so because the x part of the normal mode solution contains the function $\sin \pi n x/L$ so central to Fourier theory; all we have then done is to multiply the equation by another suitable sine function which, upon integration, has yielded a zero result except for one term in the infinite series. The sine functions we have been using have the important mathematical property called *orthogonality*, which is that the integral of the product of any two of these functions over the appropriate limits is zero unless the two functions chosen are identical. It is this orthogonality property that is basic to analyses of this kind. If, for example, the string had not been of uniform mass per unit length, the normal mode solutions would not have been sinusoidal in x and, in general, multiplication by a sine term and integration would not have solved our problem for us since there would be no orthogonality. However, sine functions are not the only orthogonal set, and functions orthogonal to the x part of the normal mode solutions can be used in the same way as the sine function in our analysis above.

EXAMPLE 9.1 The motion of stretched string with specified initial conditions
To take a specific example of initial conditions, suppose the string is initially displaced in the manner illustrated in Fig. 9.2. This profile, $y(x, 0)$ has to be described by the two equations

$$y(x, 0) = \begin{cases} \dfrac{2h}{L} x & 0 \leqslant x \leqslant \tfrac{1}{2}L, \\[4mm] 2h\left(1 - \dfrac{x}{L}\right) & \tfrac{1}{2}L \leqslant x \leqslant L. \end{cases} \tag{9.22}$$

To find the motion of the string after the midpoint has been released (at time $t = 0$), we use (9.19), in which the C_n are given by (9.21). The integral on the right-hand side of the latter equation must be expressed as two integrals, one for the region between 0 and $\tfrac{1}{2}L$ (using the first function in equations (9.22) and the other for the region between $\tfrac{1}{2}L$ and L (using the second function). From (9.21) we thus obtain

$$C_n = \frac{2}{L}\left[\int_0^{L/2} \frac{2h}{L} x \sin \frac{\pi n x}{L} \, dx + \int_{L/2}^L 2h\left(1 - \frac{x}{L}\right) \sin \frac{\pi n x}{L} \, dx\right],$$

which, after some rearrangement, becomes

$$\frac{LC_n}{4h} = \int_0^{L/2} \frac{x}{L} \sin \frac{\pi n x}{L} \, dx + \int_{L/2}^L \left(1 - \frac{x}{L}\right) \sin \frac{\pi n x}{L} \, dx.$$

Since x/L occurs frequently, we can simplify further by letting $x/L = z$, in which case $dx = L \, dz$. The last equation therefore becomes

$$\frac{C_n}{4h} = \int_0^{1/2} z \sin \pi n z \, dz + \int_{1/2}^1 (1 - z) \sin \pi n z \, dz. \tag{9.23}$$

These two integrals may be evaluated by parts. Suppose u and v are both functions of x. Then, as is well known,

$$\int_a^b u \, dv = [uv]_a^b - \int_a^b v \, du. \tag{9.24}$$

Applying (9.24) to the first integral, I_1 in (9.23), with $u = z$ and $v = \cos \pi n z$, we obtain

$$I_1 = \frac{1}{\pi^2 n^2} \sin \tfrac{1}{2}\pi n - \frac{1}{2\pi n} \cos \tfrac{1}{2}\pi n.$$

By exactly similar reasoning we evaluate the second integral, I_2, in (9.23) as

$$I_2 = \frac{1}{2\pi n} \cos \tfrac{1}{2}\pi n + \frac{1}{\pi^2 n^2} \sin \tfrac{1}{2}\pi n - \frac{1}{\pi^2 n^2} \sin \pi n.$$

Thus

$$I_1 + I_2 = \frac{2}{\pi^2 n^2} \sin \tfrac{1}{2}\pi n - \frac{1}{\pi^2 n^2} \sin \pi n$$

$$= \frac{2}{\pi^2 n^2} \sin \tfrac{1}{2}\pi n,$$

since $\sin \pi n$ is zero for all n.

Equation (9.23) finally becomes

$$C_n = 4h(I_1 + I_2)$$

$$= \frac{8h}{\pi^2 n^2} \sin \tfrac{1}{2}\pi n.$$

Thus

$$C_1 = 8h/\pi^2, \qquad C_2 = 0,$$

$$C_3 = -8h/9\pi^2, \quad C_4 = 0,$$

$$C_5 = 8h/25\pi^2, \qquad C_6 = 0,$$

and so on.

The motion of the string with the initial conditions specified by (9.22) can now be obtained by substituting for the C_n in (9.19), which gives

$$y(x, t) = \frac{8h}{\pi^2} \left(\sin \frac{\pi x}{L} \cos 2\pi f_1 t - \frac{1}{9} \sin \frac{3\pi x}{L} \cos 2\pi 3 f_1 t + \frac{1}{25} \sin \frac{5\pi x}{L} \cos 2\pi 5 f_1 t + \dots \right),$$

(9.25)

where the fundamental frequency f_1 is given by Mersenne's law as

$$f_1 = \frac{1}{2L} \sqrt{\frac{T}{\mu}}.$$

[9.15]

Since the only frequencies present in (9.25) are harmonics of the fundamental frequency f_1, the motion of the string is periodic; that is to say, the profile of the sting will assume, momentarily, its initial shape at the instants of time $1/f_1$, $2/f_1$, $3/f_1$ and so on. In practice this never happens, because the energy originally put into the string when it was drawn aside becomes dissipated owing to the viscosity of the air, and to friction effects and non-rigidity at the supports. In the case of stringed musical instruments, it is the movement of the supports, driven by that of the string, which transmits the vibration to the belly or soundboard of the instrument which, in turn, vibrates to cause the sound waves in the air.

Finally, we did not take into account the fact that all strings, to a certain extent, have rod-like characteristics (i.e. they are not perfectly flexible); the frequencies of the normal modes are only approximately given by (9.14) and they are not quite harmonically related to each other. However, despite the lack of the inclusion of all these factors in our analysis (which would have made it extremely complicated), (9.25) does provide an understanding of the principles of the vibrations of a stretched string. ∎

9.4 Formants

In the last section we analysed the transverse motion of an ideal stretched string and found that many of the properties of musical instruments could be accounted for. In this section we shall delve a little further into this topic and consider, in outline, the way in which the resonator (which may be the belly of a violin or the soundboard of a pianoforte) modifies the vibrations originating at the string. Our reason for persisting with this subject are manifold – musical sounds are in many respects the simplest kinds of all, and the physics of musical instruments, a subject worthy of study in its own right, employs principles which are applicable in many other branches of science.

The sound which we hear from a stringed musical instrument comes mainly from the cavity resonator to which the string is mechanically coupled, rather than from the string itself. This can be demonstrated quite simply by holding one end of a string of, say, one metre length and tying a weight to the lower end. When the string is plucked, very little sound emerges (although the string is under tension) because there is no suitable resonator attached to the string to cause enough of a disturbance in the air to produce an audible sound.

It is primarily the large vibrating surface area of the resonator which is responsible for the sound. Here we have a mechanical 'amplifier' which, in company with any other sort of amplifier, has different gains at different frequencies. For example, the gain of an audio amplifier (i.e. the ratio of the output voltage to the input voltage) varies with frequency, but a good amplifier will have a constant gain for most of the audio range.

Now let us consider the violin. The primary vibrators in the instrument are the strings, which generate a waveform whose spectrum (for a single note held for a long time) consists of delta functions at all the harmonics of the fundamental frequency. The various magnitudes of the delta functions are determined (within the limitations of our analysis in the last section) by the coefficients C_n in (9.19). But these are not the 'strengths' of the harmonics of the note as we hear it, because, as we have seen, it is not the string which directly generates the sound wave, but the resonator activated by the string. The 'gain' of the resonator, because it varies with frequency, will modify the relative magnitudes of the delta functions, and the spectrum of the sound eventually emitted is this modified version. The resonator is said to exert a *formant* effect upon the 'signal' it receives from the string.

To illustrate further the meaning of the formant, let us suppose that, by some means or other, we have been able to excite vibrations in the string such that all the harmonics are of equal 'strength' – that is, all the delta functions in the spectrum of the signal provided by the string are of equal magnitude. The amplitude spectrum (we will not concern ourselves with the phases of these delta functions) is sketched schematically in Fig. 9.3. Now by exciting the resonator with a pure tone of variable frequency but constant amplitude, and observing, with a microphone connected to a voltmeter, the magnitude of the sound wave emitted into the air at different frequencies, we can plot a 'gain' versus 'frequency' curve for the resonator. (Again,

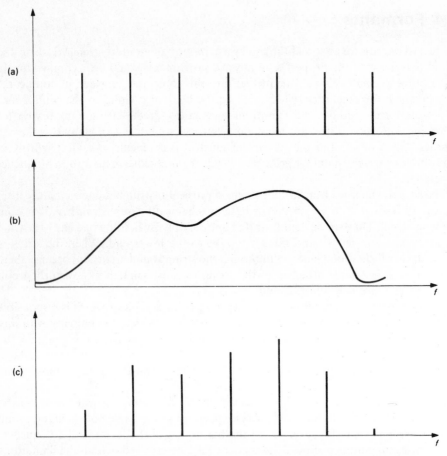

Figure 9.3 The effect of formants: (a) amplitude spectrum of signal provided by vibrations of the string; (b) amplitude formant characteristic of resonator; and (c) amplitude spectrum of sound wave emitted.

we are not concerning ourselves with the phase of the output compared with that of the input, but just with the amplitudes.) This curve is called the *formant characteristic* of the particular resonator. An example of a formant characteristic is shown schematically in Fig. 9.3b.

We are now in a position to determine the spectrum of the sound emitted. Since a pure tone of variable frequency produces output amplitudes as illustrated in Fig. 9.3b, then the signal illustrated in Fig. 9.3a will, by the principle of superposition, produce an output whose spectrum is the product of the original spectrum and the formant characteristic. This is illustrated in Fig. 9.3c. Thus, in this example, although the fifth harmonic had originally the same amplitude as all the others, it will be amplified more than the others because the formant characteristic has its maximum

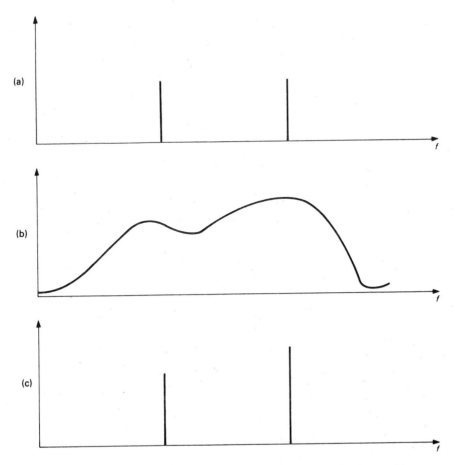

Figure 9.4 The effect of the formant of Fig. 9.3b on a tone of higher fundamental frequency: (a) amplitude spectrum of signal provided by the vibrations of the string; (b) amplitude formant characteristic of resonator; and (c) amplitude spectrum of sound wave emitted.

at this frequency So whatever the spectrum provided by the primary vibrator may be, the resonator will exert its formant effect upon it.

The extreme importance of the formant becomes very apparent when we consider the effect of playing different notes on the instrument. Suppose we continue our (admittedly rather unreal) hypothesis that the string is somehow being excited to provide a spectrum where the harmonics are of equal strength, and that we consider two different notes played by the instrument. The first note is that illustrated in Fig. 9.3, which we have already dealt with. The second, which we will assume to be higher in fundamental frequency, is illustrated in Fig. 9.4. Here, in Fig. 9.4a, we see that the harmonics are further apart from each other than in Fig. 9.3. The formant characteristic, however, is identical to that in Fig. 9.3 because it is a property of the *violin*, and has nothing to do with the note being played. As a result of the variation in position of the spectral lines but constancy of the formant characteristic, we see that the fifth harmonic is no longer the strongest; it is the *second* harmonic now that is at a frequency corresponding to the peak in the formant curve, and it is this harmonic which is therefore strongest in the emitted sound.

It is thus the formant of the belly of a violin which impresses itself upon the harmonic content of whatever the performer is playing; although the spectra of different notes are different, they all have in common that they have been multiplied by the same formant characteristic. The frequency of the fundamental tells the brains of the listeners what the pitch of the note is, and the formant characteristic, which 'colours' the spectrum of any note played, tells the listeners that they are listening all the time to a violin. Indeed, since the formant is slightly different for different violins, enough 'formant information' is presented to a sensitive musically trained listener to enable him or her to distinguish one violin from another. It is interesting to speculate to what extent the 'Kreisler' tone differs from the 'Menuhin' tone by virtue of the formant characteristics of their respective violins as opposed to their different methods of playing.

Formants are clearly very important in musical and psychological acoustics. We have dealt so far with the 'fixed' formant, i.e. an unchangeable property of a particular instrument. But formants need not necessarily be of this type. For instance, the cavities in the human head provide a formant which modifies the spectrum of sounds made by the larynx. The characteristics of this formant can be changed at will by altering the shape of the mouth cavity. By doing so we alter the spectrum of the sound emitted when we speak or sing and in this case the listener interprets the formant for a particular shape of the mouth as a *vowel*. It is evident, therefore, that the auditory system contains a very well-developed formant-recognising mechanism; exactly how this works is not yet clearly understood, but it is certain that this is one of the fundamental properties of the ear–brain combination in its capacity as an information receiver and interpreter.

9.5 The perception of sound

We have touched several times in this chapter upon the subject of perception of sound. Although the discussion has been mainly on the purely physical aspects of sound, it is impossible to isolate these entirely from the psychological aspects, because so many of the physical systems we have chosen to analyse owe their very existence to their use in musical contexts which involve sound perception. In this section we shall make a few general remarks about perception which will serve to augment what has already been said.

One of the most important facts about perception generally, and that of sound in particular, is summed up in the *Weber–Fechner law*. We can best illustrate the ideas behind this law by a simple non-acoustical example. Suppose we blindfold a suitable volunteer, and put a ten-gram weight in his hand. Then if we add another ten-gram weight on top of the first he will immediately become aware that the total weight in his hand has increased. However, if we had started by putting a kilogram weight in his hand and then added a further ten-gram weight, the increase in weight would be so slight that he would be unable to notice it. So the psychological effect of adding a given stimulus (namely the ten-gram weight) is profoundly affected by the magnitude of the stimulus which was already there.

The Weber–Fechner law is a quantitative assertion which was originally based on research into the effects of placing additional weights on the hand. It states that the increase in stimulus necessary to produce a given increase in sensation is proportional to the pre-existing stimulus. In the context of our example it means that the increase in weight that would cause our volunteer to notice that it had been increased is proportional to the weight on his hand to start with.

To put this in more mathematical form, let us suppose that the original weight was W and that the just noticeable difference in stimulus, dS, is caused by the addition of small weight dW. Then the assertion of the Weber–Fechner law is that the dW which produces this must be proportional to W. If we let the constant of proportionality connecting these quantities be k, we have

$$dS = k\,\frac{dW}{W}.$$

An integrated form of this equation is

$$S = k\,\log_e W. \tag{9.26}$$

Equation (9.26) is the mathematical description of the Weber–Fechner law as usually stated.

Now how does this apply to the perception of sound? The first and easiest example is that of the pitch of a pure tone of given frequency. The *objective* quality in the tone is the *frequency f* which corresponds to the W in (9.26); the *subjective* quality to be correlated with f is the *pitch* of the note, p, which corresponds to S in that equation. For pitch perception, therefore, (9.26) becomes

$$p = k\,\log f.$$

It is usual in this connection to define the constant k as $1200/\log 2$. Thus

$$p = \frac{1200}{\log 2}\,\log f.$$

The reason for this becomes apparent in the following way. An octave rise in pitch corresponds to a doubling of the frequency. Thus, corresponding to a frequency f_0 we have a pitch p_0 given by

$$p_0 = \frac{1200}{\log 2}\,\log f_0,$$

while the pitch p_1, corresponding to a note one octave higher than f_0, is

$$p_1 = \frac{1200}{\log 2} \log 2f_0.$$

Subtracting the two equations,

$$p_1 - p_0 = \frac{1200}{\log 2} \log \frac{2f_0}{f_0}$$

$$= 1200,$$

giving us the rise in pitch (on the scale we have defined) corresponding to one octave. Now on the equal-tempered scale, the octave is divided into 12 equal pitch increments of semitones (i.e. equal frequency ratios) namely from A to A#, A# to B, B to C, and so on up to the A above; thus one semitone becomes equivalent to a pitch increase of $1200/12 = 100$ units. The original definition of the constant k as $1200/\log 2$ therefore produces the effect of dividing each semitone into 100 parts; the unit of pitch resulting from this definition, which is widely used when discussing pitch, is called the *cent* (although other units have been devised for special purposes).

We see that the Weber–Fechner law is particularly applicable when describing musical pitch. It is not in fact true for very high and very low frequencies in the audible range, but is a very good approximation over a range of several octaves.

Let us now consider the loudness of a pure tone. Just as pitch was the subjective correlate of frequency, so is loudness the subjective correlate of intensity. Experiments have been performed to find the just noticeable increase in loudness upon increase in intensity and it has been found that the Weber–Fechner law holds reasonably well, the loudness L corresponding to the subjective quality S on the left-hand side of (9.26), and intensity I to W on the right-hand side. The equation therefore becomes

$$L = k \log I. \tag{9.27}$$

Departures from the above relationship, however, occur at very high and very low intensities. We have already mentioned the case of low intensities; below a certain *threshold* of intensity no sound can be heard at all – it therefore follows that there can be no psychological correlate of *increase* of intensity below the threshold. The threshold is a function of frequency and is at its lowest for frequencies in the range 1–2 kHz (i.e. the ear is at its most sensitive in this frequency range). At high intensities the sensation becomes that of pain, and the relationship (9.27) breaks down.

It is usual to define a scale of L such that k is unity and the logarithms are taken to the base 10. Experiment has shown that the usual root mean square pressure threshold at a frequency of 1 kHz is about 2×10^{-5} N m^{-2}. Let us designate the intensity corresponding to this pressure by I_0 and the loudness zero. We should therefore modify (9.27) to read

$$L = \log_{10} I - \log_{10} I_0 = \log_{10}(I/I_0) \tag{9.28}$$

On this scale of L, the difference in loudness between sounds 1 and 2 is $\log_{10}(I_1/I_2)$ bels, after A.G. Bell who did so much of the pioneer work in this field. This turns out to be a rather cumbersome unit as we quickly find out by means of a simple example. Suppose the intensity is initially I_1, with a corresponding loudness L_1, and is then doubled, giving a loudness L_2. Initially, therefore, (9.28) becomes

$$L_1 = \log_{10} \frac{I_1}{I_0},$$

and after doubling I_1, it becomes

$$L_2 = \log_{10} \frac{2I_1}{I_0}.$$

Subtraction of the first of these equations from the second yields

$$L_2 - L_1 = \log_{10} \frac{2I_1}{I_0} - \log_{10} \frac{I_1}{I_0}$$

$$= \log_{10} 2$$

$$\cong 0.3 \text{ bels,}$$

a rather small number for a twofold increase in intensity. In practice we therefore use a unit one-tenth the size of the bel, called the *decibel*, abbreviated in most scientific literature to dB, and in most engineering literature to db. A doubling of intensity therefore produces an increase in loudness of $10 \times 0.3 = 3$ dB. As a rule of thumb, an increase of intensity of 1 dB is just perceptible.

As we have seen, 0 dB is defined as corresponding to a root mean square pressure p_0 of 2×10^{-5} N m^{-2} at 1000 Hz. Since the intensity is proportional to the square of the acoustic pressure, (9.28) becomes

$$L = \log_{10} \frac{p^2}{p_0^2} \text{ bels}$$

$$= 10 \log_{10} \frac{p^2}{p_0^2} \text{ dB}$$

$$= 20 \log_{10} \frac{p}{p_0} \text{ dB.}$$

The right-hand side of this equation defines the *sound pressure level* (abbreviated to SPL) corresponding to the root mean square acoustic pressure p. Thus

$$\text{SPL} = 20 \log_{10} \frac{p}{p_0} \text{ dB.}$$

Although the SPL is useful in psychological work in acoustics, it must be emphasised that it is itself not a subjective quantity at all, as the last equation shows. It is a function merely of the prevailing root mean square sound pressure and the standard p_0.

The SPL is used not only to designate the sound pressure of a pure tone in a form which has relevance to its perception, but can be extended to define the level of *any* sound whose root mean square pressure p can be determined. It is still, however, defined with reference to a pure tone of 1000 Hz and of root mean square pressure p_0 of 2×10^{-5} N m^{-2}. To give some qualitative idea of the loudness corresponding to various SPLs the following list may be useful. An SPL of 15 dB corresponds roughly to the ambient noise inside a broadcasting studio, 60 dB to that in the average living room during a conversation, 100 dB to that inside an underground train, and 130 dB to the noise made by a pneumatic drill situated a few metres away. Beyond about 140 dB one feels a sensation of pain, and this value is the so-called *threshold of pain*.

EXAMPLE 9.2 Addition of two sound intensities

Suppose we have two sound sources, which are individually 90 and 100 dB. What is the loudness when they are both sounded together?

From the foregoing paragraph, the reader will appreciate that the answer is certainly not 190 dB! To obtain the correct answer, we use (9.28)

$$L = \log_{10} \frac{I}{I_0},$$

where I_0 is the intensity of the agreed origin of the decibel scale (corresponding to a root mean square pressure of 2×10^{-5} N m^{-2}).

For the first sound source then,

$$I_1 = I_0 10^{9.0},$$

and for the second source,

$$I_2 = I_0 10^{10.0}.$$

We now must assume that the sources are *incoherent*, see section 10.3.3, i.e. they produce no interference effects. This would not be the case if, for example, they were two sinusoidal tones of the same frequency; they would interfere, and the perceived intensity would depend on exactly where the observer was situated. On the other hand, if the sound sources were not correlated (for instance, a street band and a passing aeroplane), the total intensity I is the sum of the individual intensities, thus

$$I = I_1 + I_2 = I_0(1 + 0.1) \times 10^{10}$$

whence the loudness is

$$\log_{10}(I/I_0) = \log_{10}(1.1 \times 10^{10})$$

$$= 10.04 \text{ bels} = 100.4 \text{ dB.} \qquad \blacksquare$$

Problems

1. The normal modes for vibrations of the air in a pipe closed at one end were shown to have frequencies $(2n + 1)c/4L$. What is the smallest value of n for which the frequencies corresponding to n and $n + 1$ are less than a semitone apart? What is this interval in cents?

2. Deduce the frequencies of the normal modes for a pipe of length L open at both ends.

3. Use a calculator to deduce $(3/2)^n$ for $n = 2, 3, 4, \ldots$ At what value of n is this very close to an integral power of 2? Of what musical significance is this?
    Show that $(3/2)^4$ is very close to $2^2 \times 5/4$. Of what musical significance is this?

4. By considering the pressure wave in a closed organ pipe, deduce whether the end correction would cause the pipe to sound at a higher or a lower pitch than that which would apply in the absence of an end correction.

5. A string of mass 0.01 kg is stretched between two points 0.5 m apart. The tension in the string is 200 N. What is the frequency of (i) the fundamental and (ii) the third harmonic?

6. A string is stretched between two points, and has a certain fundamental frequency. If these two points are moved slightly further apart, then T, μ and L all change; show that the fundamental frequency must necessarily increase.

7. A uniformly stretched string of linear density μ and length L under a tension T is initially displaced so that the points one-quarter of the way from each end are oppositely displaced through a distance h, leaving the centre point unmoved. The string is released from this position at time $t = 0$.
    Assuming that $h \ll L$, deduce the nature of the subsequent motion of the string. What harmonics are missing?

8. Deduce the energy in each of the harmonics for the vibrating string considered in Example 9.1.

9. The sound level at a distance 100 m from an aeroplane is 100 dB. Assuming that the intensity is inversely proportional to the square of the distance away, deduce the distance at which the sound would be just audible.
    Explain why this answer is absurd.

10. Deduce an expression for the pressure variations representing a pure tone of peak acoustic pressure p_1, of duration 5 s, the pitch rising at a constant rate from an initial value of 200 Hz to a final value of 3200 Hz. (Hint: refer to section 8.4.2.)

11. Two pure tones, 0.1 Hz apart in frequency, sound simultaneously in an acoustically dead room. A sound-level meter in the room records 60 dB as its maximum reading, and 20 dB as minimum. If 0 dB corresponds to a root mean square acoustic pressure of 2×10^{-5} N m^{-2}, calculate the acoustic pressure amplitudes of each of the two pure tones.

12. Two independent sources of sound have loudnesses 50 and 55 dB when sounded separately. What is the loudness when they are sounded together?

13. A sinusoidal voltage source is connected to the input of an amplifier, and a loudspeaker to its output. If the input voltage is doubled, by how many dB is the output increased?

14. In many receivers and amplifiers, the volume control is 'logarithmic', i.e. the resistance between B and C is not proportional to the angle through which the control is rotated (as with a 'linear' control), but rather as shown below. Why is this a good arrangement?

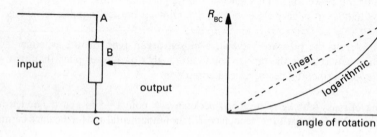

10 Electromagnetic waves

10.1 The electromagnetic theory of light

Although the wave nature of light had been established in the early part of the nineteenth century, when Thomas Young performed his classic interference experiment, it took a further half century before a satisfactory theory for light waves emerged. The great theoretical difficulty which had to be overcome was the explanation of how waves could be propagated in the absence of a medium, for all waves known at that time were mechanical in type and could not exist in the absence of a material medium.

The electromagnetic theory of light was formulated by James Clerk Maxwell in 1860. Maxwell had developed a set of equations, now referred to universally as Maxwell's electromagnetic equations, which related electric and magnetic quantities, and, in fact, summarised all the known properties of electric and magnetic fields. Maxwell showed that these equations could be processed in such a way as to yield a wave equation. He went on to show from this that electromagnetic waves could be propagated in free space, with a wave velocity which could be deduced from electrical measurements unconnected with wave phenomena, which turned out to be very nearly the same as the velocity of light. The latter velocity had been approximately determined, over a hundred years earlier, by the astronomers Römer and Bradley, and later (more precisely) by Fizeau and Foucault.

We shall not reproduce the arguments leading up to Maxwell's equations and the wave equation, as we did in the first edition. These can be found in all the standard texts on electromagnetism (see for example *Electricity and Magnetism* by W.J. Duffin, 3rd edn, McGraw-Hill, 1980; or *Electromagnetism* by I.S. Grant and W.R. Phillips, 2nd edn, Wiley, 1990). Instead we merely state, in the form explained at the end of section 4.4.3, the partial differential equations for waves in free space which emerge from Maxwell's equations, viz.:

$$\nabla^2 \mathbf{E} = \mu_0 \varepsilon_0 \frac{\partial^2 \mathbf{E}}{\partial t^2} \tag{10.1}$$

$$\nabla^2 \mathbf{B} = \mu_0 \varepsilon_0 \frac{\partial^2 \mathbf{B}}{\partial t^2} \tag{10.2}$$

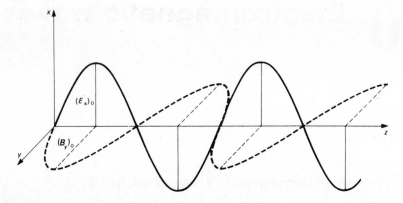

Figure 10.1 Perspective representation of a plane harmonic electromagnetic wave.

Here $\mathbf{E}$ is the electric field intensity, $\mathbf{B}$ the magnetic field flux density (both vectors) and μ_0 and ε_0 are respectively the permeability and permittivity of free space. The quantity $\mu_0 = 4\pi \times 10^{-7}$ H m^{-1} and ε_0 can be measured, giving a phase velocity

$$c = \frac{1}{\sqrt{(\mu_0 \varepsilon_0)}} \tag{10.3}$$

which agrees with the measured speed of light (3×10^8 m s^{-1}).

Equations (10.1) and (10.2) have harmonic plane-wave solutions

$$\mathbf{E} = \mathbf{E}_0 \exp i(\mathbf{k\cdot r} - \omega t) \tag{10.4}$$

$$\mathbf{B} = \mathbf{B}_0 \exp i(\mathbf{k\cdot r} - \omega t). \tag{10.5}$$

Both of these are needed for a complete description of the wave – the changing B field is always accompanied by a changing E field and vice versa. The wave is entirely transverse and also $B = E/c$. If the E disturbance is in one direction only, we have a *plane-polarised* wave. It turns out that B is perpendicular to E, and the vectors $\mathbf{E}$, $\mathbf{B}$ and $\mathbf{k}$ are mutually orthogonal and form a right-handed system taken in this order. Figure 10.1 shows such a plane-polarised wave in which E is along Ox, B along Oy so that the direction of propagation is along Oz. We may write the wave equations for this situation in terms of the scalar components as follows:

$$E_x = (E_x)_0 \exp i(kz - \omega t) \tag{10.6}$$

$$B_y = (B_y)_0 \exp i(kz - \omega t). \tag{10.7}$$

Shortly after the appearance of Maxwell's theory of electromagnetic waves, Hertz provided experimental corroboration in which electromagnetic disturbances produced by an oscillating circuit were received by an isolated detector. Following the establishment of Maxwell's theory, a wide range of different types of radiation became

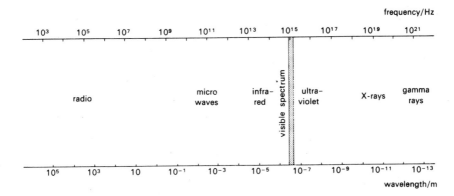

Figure 10.2 The electromagnetic spectrum (frequency and wavelength plotted on a logarithmic scale).

recognised as electromagnetic waves of different frequencies (and therefore wavelengths). Radio waves, microwaves, infrared radiation, visible light, ultraviolet rays, X-rays and γ-rays are all examples of electromagnetic waves, differing from one another only in wavelength. They form the *electromagnetic spectrum*, which is illustrated in Fig. 10.2. The wavelength range covered by the spectrum is enormous – from hundreds of miles for radio waves originating from distant nebulae down to one ten-thousand-millionth part of a millimetre for γ-rays. We see from Fig. 10.2 that light waves – the only electromagnetic waves which stimulate the eye – cover only a tiny part of this spectrum.

So far we have only considered electromagnetic waves in free space. Equations (10.1) and (10.2) can be extended to cover a perfectly insulating, homogeneous, isotropic medium merely by replacing ε_0 and μ_0 by $\varepsilon_r\varepsilon_0$ and $\mu_r\mu_0$, where ε_r and μ_r are numerical (dimensionless) constants, known respectively as the *relative permittivity* and the *relative permeability* of the medium. Such a medium is referred to as a *dielectric*; we are particularly interested here in dielectrics in which light waves may be propagated. Thus

$$\nabla^2 \mathbf{E} = \varepsilon_r\varepsilon_0\mu_r\mu_0 \frac{\partial^2 \mathbf{E}}{\partial t^2}, \tag{10.8}$$

$$\nabla^2 \mathbf{B} = \varepsilon_r\varepsilon_0\mu_r\mu_0 \frac{\partial^2 \mathbf{B}}{\partial t^2}. \tag{10.9}$$

We conclude from (10.8) and (10.9) that electromagnetic waves can be propagated in a dielectric with wave velocity

$$v = \frac{1}{\sqrt{(\varepsilon_r\varepsilon_0\mu_r\mu_0)}}. \tag{10.10}$$

It turns out that ε_r cannot be less than unity; and in the case of a dielectric, μ_r is approximately unity. Thus the velocities of electromagnetic waves in media are less than the velocity in free space.

If we reserve the symbol c for the wave velocity in free space, we see from (10.10) that

$$v = \frac{c}{\sqrt{(\varepsilon_r \mu_r)}}. \qquad (10.11)$$

The *refractive index n* of a dielectric medium is defined as the ratio of the wave velocity in free space to that in the medium, i.e.

$$n = \frac{c}{v}; \qquad (10.12)$$

we see therefore from (10.11) that $n = \sqrt{(\varepsilon_r \mu_r)}$. Since $\mu_r \cong 1$,

$$n \cong \sqrt{\varepsilon_r}. \qquad (10.13)$$

Equation (10.13) would seem to indicate that the refractive index of a dielectric is constant. The quantity ε_r, however, turns out to be frequency dependent. If the value of ε_r measured by a statical method is used, then (10.13) holds reasonably well at low frequencies but breaks down completely at high frequencies.

10.2 Polarisation

It was noted above that plane light waves (indeed plane electromagnetic waves in general) are transverse. The experimental evidence for the transverse nature of light waves is the fact that light waves can be *polarised*. Light beams obtained from most common sources (e.g. the hot gas in a discharge tube) are made up of a very large number of waves. The **E**-vectors of these waves (we are still assuming plane waves) are arranged in all directions in space normal to the propagation direction, and such light beams are said to be *unpolarised*. If the light is filtered in some way such that the **E**-vectors are suppressed in all directions save one, we are left with a unique direction of vibration and the light is said to be *plane-polarised*.

Light may be made plane-polarised in one of a number of ways. One method is by passing the light through a Nicol prism, that is, a pair of calcite crystals cut in a special way and cemented together. Another way is by passing light through a dichroic crystal – one for which the absorption of light is much greater for a given direction of vibration of **E** than for the direction perpendicular to it, so that if the light passes through a sufficient length of such a crystal, effectively all the amplitude in one direction is removed. This is the basis of the Polaroid method. A third method is by reflection; when a beam of light is directed onto a plane glass surface at an angle of incidence of about 57° (Brewster's angle), none of the light with electric vector vibrating parallel to the plane of reflection is reflected, so the reflected light is plane-polarised, with vibration direction perpendicular to the reflection plane.

If a beam of light is plane-polarised by, say, passing it through a Nicol prism, and then passed through a second Nicol prism which can be rotated about the direction of propagation of the light, it is found that for certain angular positions of the second prism no light emerges from it. This is incontrovertible experimental evidence that light waves are transverse. The first prism is referred to as the *polariser* and the second as the *analyser*. If the polariser and analyser are rotatable and are equipped with angular scales, this simple system can be used to investigate the state of linear polarisation of any light beam. An application of this is in crystal physics, for many crystals polarise light, and analysis of the state of the light emerging from such crystals can yield important information.

Light may also be made *circularly* polarised. This requires a quarter wave plate cut from a *birefringent* crystal. A birefringent crystal is anisotropic (different physical properties in different directions). A consequence of this is that light waves travel through the crystal at different velocities depending on the angle the **E**-vector in the incident wave makes with a unique direction in the crystal known as the optic axis. Quartz and calcite are two examples. We define two refractive indices: n_e, the extraordinary refractive index, for light with **E** parallel to the optic axis; and n_o, the ordinary refractive index, for light with **E** perpendicular to the optic axis. If plane-polarised light is incident onto a slab of birefringent crystal with **E** inclined at 45° to the optic axis as shown in Fig. 10.3, we can resolve this into two waves of equal amplitude having their electric vectors respectively parallel to the Ox and Oy-axes. The optic axis is along Ox. As the crystal is birefringent, the two component waves will traverse it at different speeds. If for this crystal $n_e > n_o$ as in quartz, the Ox wave will travel slower than the Oy wave (since $n = c/v$) and will therefore lag in phase. Suppose the thickness of the crystal is such that the emerging waves differ in phase by $\pi/2$ exactly. (Such a crystal is known as a quarter wave plate because a phase change of $\pi/2$ corresponds to a path difference of a quarter wavelength.) What sort of resultant emerging wave would we get? We already know what happens from section 2.2.2 where we compounded two equal s.h.m.s at right angles differing in phase by $\pi/2$. The resultant path was a circle. In this case the tip of the resultant **E** vector describes a circle when viewed down the Oz-axis, but the wave itself is a spiral as shown in Fig. 10.3. This is circularly polarised light. Circularly polarised light is not immediately distinguishable from unpolarised light. For example, if circularly polarised light is examined by rotating a single piece of Polaroid, uniform but reduced intensity will be seen at all angles as would be found for unpolarised light examined in the same way. Circularly polarised light can, however, be identified with the aid of a quarter wave plate. If the circularly polarised light is incident normally on a quarter wave plate, a further $\pi/2$ phase difference is produced between the component waves so the emergent waves are π out of phase. The resultant wave is therefore plane-polarised and can readily be distinguished as such with a piece of Polaroid.

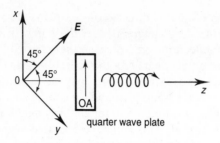

Figure 10.3 Production of circularly polarised light. The quarter wave plate has its optical axis OA parallel to Ox.

10.3 Interference and diffraction

10.3.1 Introduction: Huygens' principle and the scalar approximation

These are subjects of very great importance and interest which really require a book to themselves. It is impossible here to do more than merely introduce them in a rather general way, and to show their significance within the wider subject of wave motion.

Interference takes place when light waves are superposed. Diffraction results when light waves are impeded or restricted in some way, for example, by placing an opaque obstacle in their path. It is enormously difficult to determine the effect of an obstacle on light waves from first principles. We should have first to find the electric and magnetic boundary conditions placed upon (10.1) and (10.2) by the presence of the obstacle, and then to solve the equations according to these conditions. Fortunately, results, correct to a very good aproximation, can be obtained in a very much simpler way by means of the *scalar approximation* and *Huygens' principle.*

According to the scalar approximation, the disturbance caused by a light wave may be represented by a single scalar variable when interference and diffraction effects are being considered (but not, obviously, for polarisation). So with this approximation we can represent a plane harmonic light wave by

$$\phi = a \sin(\omega t - kx),$$

where ϕ is the scalar disturbance.

Huygens' principle states that each point on a wavefront may be regarded as a new source of waves. It is relatively easy to show that Huygens' principle is consistent with the laws of rectilinear propagation, reflection and refraction. This we will now do, starting with rectilinear propagation. A plane wavefront AB (Fig. 10.4) is moving in the direction of the arrow with velocity v; at any instant we may regard each point on AB as a source of secondary wavelets which give rise to spherical wavefronts. After a time t these wavefronts will all have radius vt and their common tangent A'B' will clearly occupy the same position as that which the original wavefront will have

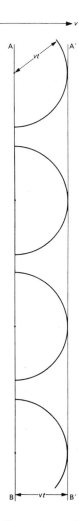

Figure 10.4 Rectilinear propagation of a plane wavefront.

reached after time t. Thus Huygens' principle is consistent with rectilinear propagation, as long as we choose to ignore the wave moving in the reverse direction which Huygens' principle predicts.

We will examine next the law of reflection. In Fig. 10.5 a plane wavefront AB is shown incident at an angle of incidence i upon a plane reflecting surface XY at the instant when the point A has just reached the surface. We can regard A as a Huygens point source. In the further time it takes for B to reach the reflecting surface at D, the radius of the spherical wavefront emanating from A will grow to the value AC; since the incident and reflected waves are in the same medium, they will have the same velocity, so that AC and BD are equal. The reflected wavefront is obtained by

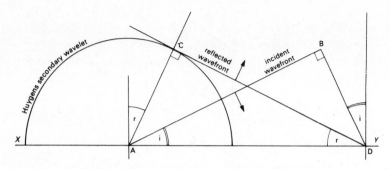

Figure 10.5 Reflection of a plane wavefront.

drawing the tangent from D to the surface of the spherical wavefront at C, so that the normal to CD gives the direction of travel of the reflected wave. We see at once that the triangles ACD and ABD are congruent; the angles BAD and CDA are therefore equal and the angle of incidence is equal to the angle of reflection.

A similar construction (Fig. 10.6) leads to the law of refraction. Here PQ represents the plane boundary between two media of refractive indices n_1 and n_2 respectively ($n_2 > n_1$). Again we have a plane wavefront AB incident at angle of incidence i on to the boundary and shown at the instant when the point A has just reached the boundary. Suppose the point B takes a further time t to reach the boundary at C, so that $BC = v_1 t$ (where v_1 is the velocity of light in the first medium). To find where the disturbance due to A has reached in this time, we regard A as a Huygens secondary source and draw a sphere of radius $v_2 t$ with centre A (v_2 is the velocity in the second medium). The wavefront in the second medium is represented by CD, the tangent

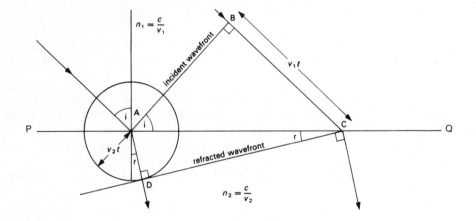

Figure 10.6 Refraction of a plane wavefront.

from C to the section through the sphere in the plane of the drawing. The angle between the normal to CD and the normal to the boundary is the angle of refraction r. We note from the definition of refractive index (10.12) that, since $n_2 > n_1$, the velocity must be less in the second medium than in the first. From the geometry of Fig. 10.6 we see that

$$\sin i = \frac{BC}{AC} = \frac{v_1 t}{AC}$$

and

$$\sin r = \frac{AD}{AC} = \frac{v_2 t}{AC},$$

therefore

$$\frac{\sin i}{\sin r} = \frac{v_1 t}{v_2 t} = \frac{c/n_1}{c/n_2},$$

from (10.12), so that

$$\frac{\sin i}{\sin r} = \frac{n_2}{n_1}.$$

Thus the sines of the angles of refraction and incidence are in the same ratio as the refractive indices of the media on either side of the boundary and this ratio is constant for a given wavelength; this is the law of refraction or *Snell's law*.

10.3.2 Interference

We have established that Huygens' principle is reasonable by showing that the three fundamental laws of geometrical optics can be deduced from it; we now go on to apply the principle to try to account for the phenomena of interference and diffraction. Let us first investigate what happens when an observer receives light waves *simultaneously* from two point sources. Let the point sources S_1 and S_2 (Fig. 10.7) be a distance a apart, and let the observer be at a point P at distances x_1 and x_2 from S_1 and S_2 respectively. Let us represent the disturbances at P due to the two light waves as

$$\phi_1 = a_1 \sin(\omega t - kx_1), \tag{10.14}$$

$$\phi_2 = a_2 \sin(\omega t - kx_2 - \varepsilon). \tag{10.15}$$

Here, we are assuming that we have pure sine waves, both sources emitting with the same frequency and wavelength but different amplitudes. The phase constant ε in (10.15) takes care of any difference in phase between the sources themselves. By the

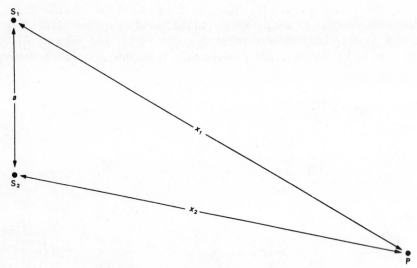

Figure 10.7 Light waves from two point sources.

principle of superposition, the net disturbance ϕ at P is the sum of the individual disturbances ϕ_1 and ϕ_2. So we have

$$\phi = \phi_1 + \phi_2 = a_1 \sin(\omega t - kx_1) + a_2 \sin(\omega t - kx_2 - \varepsilon). \qquad (10.16)$$

For a fixed position P, the only quantity on the right-hand side which varies is t, so for convenience we will write

$$kx_1 = \alpha_1, \qquad (10.17)$$

$$kx_2 + \varepsilon = \alpha_2.$$

When we substitute these into (10.16) we get

$$\phi = a_1 \sin(\omega t - \alpha_1) + a_2 \sin(\omega t - \alpha_2).$$

When we expand the sine terms and factorise, this becomes

$$\phi = (a_1 \cos \alpha_1 + a_2 \cos \alpha_2) \sin \omega t - (a_1 \sin \alpha_1 + a_2 \sin \alpha_2) \cos \omega t.$$

If we now define two further quantities R and θ such that

$$R \cos \theta = a_1 \cos \alpha_1 + a_2 \cos \alpha_2,$$

$$R \sin \theta = a_1 \sin \alpha_1 + a_2 \sin \alpha_2,$$

and then substitute these into the previous equation, we obtain

$$\phi = R \sin(\omega t - \theta).$$

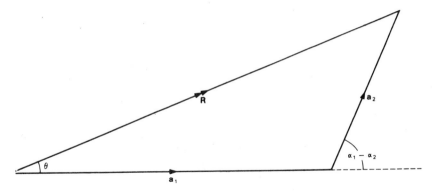

Figure 10.8 Vector addition of sine waves.

We see, therefore, that the net disturbance at P is simple harmonic in character, with the same frequency as that of the original waves, and of a phase angle relative to them defined by θ. It will be seen from squaring and adding $R \cos \theta$ and $R \sin \theta$ that

$$R^2 = a_1^2 + a_2^2 + 2a_1 a_2 \cos(\alpha_1 - \alpha_2) \tag{10.18}$$

and also that

$$\tan \theta = \frac{a_1 \sin \alpha_1 + a_2 \sin \alpha_2}{a_1 \cos \alpha_1 + a_2 \cos \alpha_2}.$$

It is worth noting, in passing, that (10.18) has the familiar appearance of the cosine formula for the solution of triangles. This suggests a useful graphical method for 'adding' two sine waves of the same frequency. If we represent the waves by two vectors whose lengths are proportional to the respective amplitudes of the waves, and the angle between the vectors is the difference between the phases of the waves, as shown in Fig. 10.8, then the amplitude of the resultant is obtained (on the same scale) by completing the triangle, and the phase of the resultant with respect to the first wave is given by the angle between the respective vectors.

In the case of light waves, it is the intensity, not the amplitude, which is the important quantity. Detectors of light waves, such as the eye and photographic film, all respond to intensity. Indeed, there is no known method of measuring light amplitude directly. We saw in section 6.3 that intensity is proportional to the square of the amplitude, for waves in strings. This is also true for light waves, so we may take R^2 (10.18) as a measure of the way in which the intensity due to the superposition of our two light waves varies from place to place as we move P around. (Strictly, as P moves, a_1 and a_2 will vary, since we are dealing with spherical waves; we will, however, ignore this variation.) For simplicity we will put $a_1 = a_2 = A$, and assume that the two sources S_1 and S_2 are exactly in phase with each other, so that $\varepsilon = 0$. When we make these simplifications (10.18) becomes

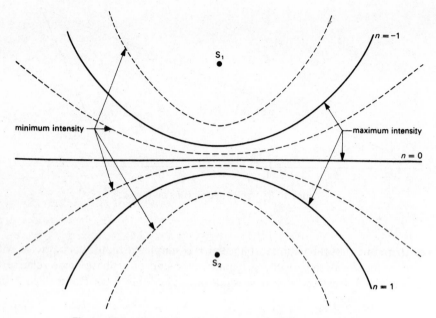

Figure 10.9 Hyperbolas of maximum and minimum intensities.

$$R^2 = 2A^2[1 + \cos(\alpha_1 - \alpha_2)].$$

Thus R^2 is maximum when

$$\cos(\alpha_1 - \alpha_2) = +1, \tag{10.19}$$

i.e. when

$$\alpha_1 - \alpha_2 = 2n\pi;$$

and minimum (zero) when

$$\cos(\alpha_1 - \alpha_2) = -1,$$

i.e. when

$$\alpha_1 - \alpha_2 = (2n + 1)\pi. \tag{10.20}$$

In (10.19) and (10.20), n is zero or any positive or negative integer. But we see from (10.17) that

$$\alpha_1 - \alpha_2 = k(x_1 - x_2), \tag{10.21}$$

since we have set ε to be zero. So the conditions (10.19) and (10.20) become, on replacing k by $2\pi/\lambda$

$$x_1 - x_2 = n\lambda \qquad \text{for maxima}$$

and

$$x_1 - x_2 = (n + \tfrac{1}{2})\lambda \quad \text{for minima.}$$

Thus, if P moves from place to place, we shall have maximum intensity ($R^2 = 4A^2$) when

$$S_1P - S_2P = n\lambda,$$

and minimum intensity ($R^2 = 0$) when

$$S_1P - S_2P = (n + \tfrac{1}{2})\lambda.$$

If S_1, S_2 and P are always in the same plane, then the curves of maximum and minimum intensity form a family of hyperbolas, as shown in Fig. 10.9.

One might suppose, on the basis of the above reasoning, that two point sources of light would produce a pattern of bright and dark lines of the kind shown in Fig. 10.9. However, if completely separate sources are used, we do *not* get such a pattern. The reason for this will now be considered.

10.3.3 Coherence

In the preceding sections, it has been assumed that the sources S_1 and S_2 emit pure sinusoidal waves of the same frequency. At any point P where the waves superimpose, there will be a summation of two waves as given by (10.16). They need not be of the same amplitude, but if they have the same frequency, there will be a *phase difference* between them at P which is *constant in time*. In any situation where this is true, the sources are said to be *coherent*. An example of coherent acoustic sources is that of two loudspeakers S_1 and S_2 connected to a common alternating voltage source.

One way of producing two coherent sources of light would be to illuminate two pinholes S_1 and S_2 with a laser beam (Fig. 10.10a), an important feature of a laser being that it produces radiation which, for our purposes at least, is monochromatic.

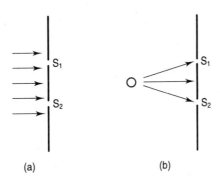

(a)   (b)

Figure 10.10 Illumination of two slits by (a) a laser and (b) a sodium lamp.

Now suppose we were to use a sodium lamp instead of the laser (Fig. 10.10b). Since a sodium lamp is a good approximation to a monochromatic source one might expect S_1 and S_2 to be illuminated coherently. However, this is not the case. Any attempt to produce an interference pattern in the region to the right of S_1 and S_2 would result in failure. This is not because the sodium lamp produces two different wavelengths in the yellow part of the spectrum (as well as many other weaker spectral lines). Even if a filter were inserted to pass just one of the yellow sodium lines, the sources would still be incoherent. To understand why, we now consider the mechanism whereby light is produced in the sodium lamp. There is an assemblage of sodium atoms all moving about with random velocities. When collisions take place, the energy acquired in the collision process by one of the atoms may cause an electron in it to be excited into a higher energy state. Subsequently, the electron in the atom returns to its original state and in doing so releases a burst of light energy. This light energy is in the form of a light wave of very limited duration – of the order 10^{-8} s (i.e. a complete wave resulting from this process would take only 10^{-8} s to pass a point). It follows from the velocity of light in free space ($\sim 3 \times 10^8$ m s^{-1}) that the light waves would extend over about 3 m in space. The frequency f of such a light wave is related to the difference in energy ΔE between the two energy states of the atom by

$$\Delta E = hf, \tag{10.22}$$

where h is the Planck constant. But we have seen in Chapter 7 that the only wave which has a single frequency is a sinusoidal wave of infinite duration, so the present short-duration wave will have a spread of frequencies whose mean will be the f of (10.22). The wavelength spread of spectral lines resulting from this effect, though measurable, is extremely small. Actually, a considerable contribution to the broadening of spectral lines is due to the Doppler effect, since the atomic sources are moving randomly.

An extremely important feature of these waves of atomic origin is that when the various atoms emit waves, their phases are randomly related. It follows therefore that two separate light sources, each consisting of many millions of individual atomic sources, cannot possibly have a constant phase relationship for times which are more than a fraction of 10^{-8} s.

We now return to the discussion in section 10.3.2. If we have two incoherent sources, the resultant intensity is still given by (10.18) – except that the term $\cos(\alpha_1 - \alpha_2)$ is not constant, as it would be for coherent sources, but varies very rapidly with time. The average of this quantity over a long time is zero, so the resultant is

$$R^2 = a_1^2 + a_2^2.$$

That is, for incoherent sources, the resultant is everywhere just the sum of the individual intensities, and consequently there is no pattern of maximum and minimum intensity of the kind shown in Fig. 10.9.

The simplest way to overcome this problem of incoherence when using a spectral lamp is to derive both sources from a common original source. One way of doing this is shown in Fig. 10.11, in which the light waves emanating from a point source

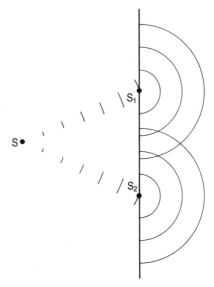

Figure 10.11   Coherent sources derived from a single common source S.

S fall onto two pinholes S_1 and S_2 in an otherwise opaque screen. By Huygens' principle, S_1 and S_2 act as sources of secondary spherical wavelets. In the region to the right of the screen, where these secondary wavelets superpose, we have the necessary condition to observe intensity variation, for a constant phase relationship between S_1 and S_2 is maintained. The pattern of intensity we see is known as an interference pattern, and the process which produces this is known as optical interference. At any point in the region where the interference pattern is clearly seen, we deduce that a disturbance arrives via S_1 at the same instant as another disturbance arrived via S_2, but that these two disturbances started off from S at times which differed from one another by only a small amount compared with 10^{-8} s.

In practice, it is much more convenient to use slits rather than pinholes for S, S_1 and S_2 in Fig. 10.11. This arrangement is known as Young's slits after Thomas Young who first performed the experiment in 1807. If monochromatic light emerging from S_1 and S_2 is viewed at a distance D from the screen, which is very many times greater than the separation a of S_1 and S_2, then a series of alternate bright and dark fringes (Young's fringes) may be seen. It is relatively easy to show, though we will not do so here, that the fringes are equispaced and that the centres of adjacent bright fringes are a distance $\lambda D/a$ apart. Thus, if we wish to obtain distinct fringes of separation, say, half a millimetre with light of wavelength 5×10^{-7} m, we must arrange for the ratio of D/a to be 1000:1.

Young's fringes are an example of interference by *division of wavefront*, since the slits S_1 and S_2 select different portions of a wavefront in order to provide Huygens' secondary sources. Other common examples of interference by division of wavefront are Lloyd's mirror and Fresnel's biprism.

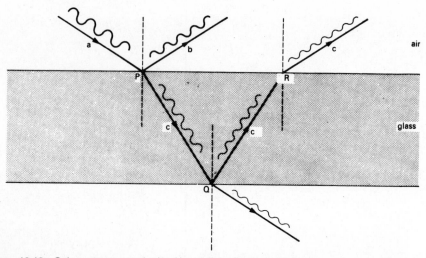

Figure 10.12 Coherent sources obtained by partial reflection of light waves at an air–glass interface. A light wave a is incident at point P. A part b of the wave is reflected, and a part c transmitted. After internal reflection at Q, c eventually emerges at R; c is coherent with b but will, in general, have a different phase since it has traversed an extra path.

Another way of providing coherent sources is by making use of the fact that when light waves are incident upon a glass surface, part of the amplitude is transmitted through the glass and part is reflected. This is schematically shown in Fig. 10.12. The physical situation at the interface is exactly analogous to that at the junction of strings of different densities discussed in Chapter 5. The two waves b and c emerging from the glass surface fulfil the conditions necessary for interference provided the time taken for c to traverse the extra path is small compared with 10^{-8} s. If these waves are brought into superposition, by means of a lens, we shall have brightness if the respective disturbances are in phase, but darkness if they are in antiphase. This is the basis of interference by *division of amplitude*, of which Newton's rings are the best-known example.

10.3.4 Diffraction

As we mentioned at the beginning of section 10.3, diffraction occurs when waves are limited or obstructed in some way. We shall consider the case of plane, monochromatic light waves incident normally upon a parallel-sided opening in an otherwise opaque screen. The experimental arrangement for this is shown in Fig. 10.13. The plane wavefronts are derived from a lamp L with the aid of two converging lenses L_1 and L_2 and a restricting slit S_1. After passing through the diffracting slit in the screen S_2, the light waves are collected by a converging lens L_3. A screen arranged in the focal plane of L_3 will be seen to be illuminated by a series of bright and dark fringes; the

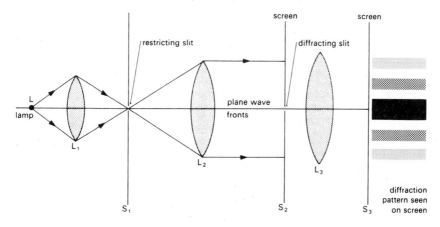

Figure 10.13 Experimental arrangement for obtaining single-slit diffraction pattern.

central fringe which is situated on the optical axis of the system will be brightest, with successive fringes above and below becoming rapidly fainter as we move away from the centre. This arrangement of fringes is known as a diffraction pattern. It is entirely due to the restriction placed on the light waves by the slit in S_2, for if S_2 is removed, it will be seen that the diffraction pattern vanishes, being replaced by a single bright line which is the optical image of the illuminated slit S_1.

We will now see how this pattern can be accounted for. Figure 10.14a shows a given plane wavefront proceeding away from the slit of width a in the screen S_2. Suppose the direction of motion of this particular wavefront makes an angle θ with the optical axis of the system. The action of the lens L_3 is to collect together all the light along this wavefront and bring it to focus along a line on the screen S_3, indicated by P_θ in the section shown in Fig. 10.14b. The wavefronts for which θ is zero are focused at P_0 on the optical axis of the system. The total effect at the point P_θ is obtained by dividing the wavefront shown in Fig. 10.14a into thin strips of equal width, the lengths of the strips being parallel to the length of the slit (i.e. perpendicular to the plane of the paper). Each strip will give rise to the same disturbance at P_θ, so we have to superpose the disturbances due to all the strips by the principle of superposition, taking account of any difference in phase between the individual disturbances. Now we note from equation (10.21) that

$$\text{phase difference} = \frac{2\pi}{\lambda} \text{ (path difference).}$$

It is apparent from Fig. 10.14a that the difference between the paths travelled by the strips at the two extreme ends of the wavefront shown is just $a \sin \theta$. Thus the extreme phase difference ϕ is given by

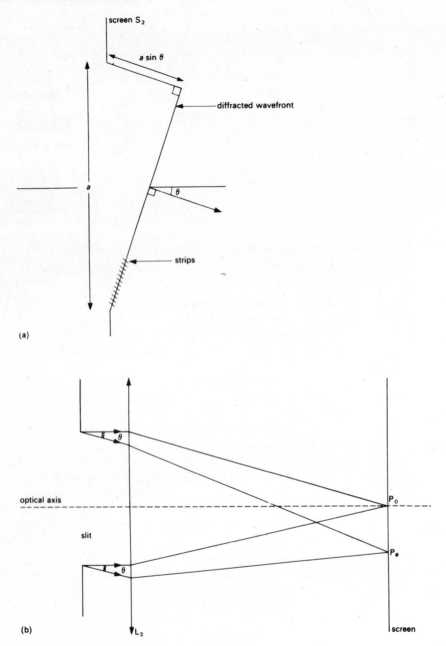

Figure 10.14 (a) Wavefront diffracted through angle θ. (b) Action of a lens in focusing wavefronts with different directions onto different positions on screen.

$$\phi = \frac{2\pi}{\lambda} a \sin \theta. \tag{10.23}$$

Therefore the extreme phase difference across the wavefront is proportional to the distance apart of the strips concerned. It follows that the phase difference between disturbances due to any two strips is proportional to their distance apart, so that since the strips are all of equal width, the difference in phase between any two adjacent strips will have a constant value which we shall designate by δ. Let there be N strips and let the amplitude at P_θ due to each be b. We can find the net disturbance at P_θ by an extension of the vector method for superposing waves illustrated in Fig. 10.8. Since we have a number N of disturbances to superpose, we have a polygon in place of a triangle, and the resultant is obtained by completing the polygon as shown in Fig. 10.15. The vectors all have the same length b, and each is inclined to its neighbour by the phase angle δ. We now let the strips become infinitely thin, and the number of them, N, infinitely large in such a way that the product Nb remains finite. As we go to this limit the vector polygon of Fig. 10.15a becomes an arc of a circle, and the resultant R is the chord to this arc as shown in Fig. 10.15b. Let the arc have centre O and radius r. Clearly the extreme phase difference ϕ (see (10.23)) is represented by the angle between the tangents to the two ends of the arc as shown.

We see from the geometry of Fig. 10.15b that

$$R = PQ = 2r \sin \tfrac{1}{2}\phi$$

and that arc $PQ = r\phi = Nb$. Thus

$$R = \frac{2Nb}{\phi} \sin \tfrac{1}{2}\phi.$$

But since $\tfrac{1}{2}\phi = (\pi/\lambda)a \sin \theta$ (from (10.23)), we have, putting $Nb = A$,

$$R = \frac{A \, \sin[(\pi a \, \sin \, \theta)/\lambda]}{(\pi a \, \sin \, \theta)/\lambda}. \tag{10.24}$$

It turns out that the diffraction pattern is only visible over a region of the screen corresponding to small values of θ, for which $\sin \theta \cong \theta$, so we may rewrite (10.24) as

$$R = \frac{A \, \sin(\pi a\theta/\lambda)}{\pi a\theta/\lambda} \tag{10.25}$$

or

$$R = \frac{A \, \sin \, \alpha}{\alpha} \tag{10.26}$$

if we substitute α for $\pi a\theta/\lambda$.

Equation (10.26) tells us how the amplitude varies from place to place across the screen. The function has the form shown in Fig. 10.16a, but since we are able to see only the *intensity* distribution, the curve shown in Fig. 10.16b of the square of (10.26), i.e.

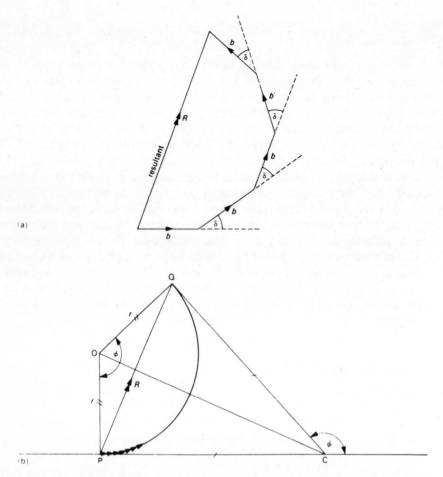

Figure 10.15 (a) Vector polygon for finding the resultant of a number of disturbances. (b) Vector diagram when individual disturbances become infinitely small.

$$R^2 = \frac{A^2 \sin^2 \alpha}{\alpha^2},$$

is of more interest. Both curves are symmetrical about $\alpha = 0$, where the zero-order maximum occurs. In the lower curve the successive maxima fall quickly away in agreement with the experimental fact mentioned earlier. In fact the zero, first, second and third maxima in Fig. 10.16b can be shown to be in the ratio $1 : \frac{1}{21} : \frac{1}{61} : \frac{1}{120}$.

It is also of interest to see the effect upon the pattern of altering the width of the diffracting slit. Now the nth maximum from the centre, at angular position θ_n, is defined by a particular value of α which we will call α_n. Since

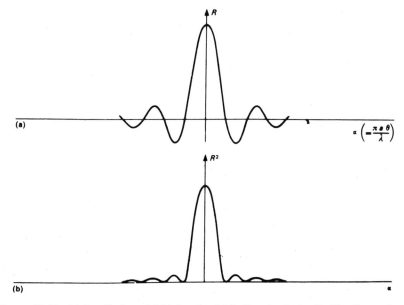

Figure 10.16 (a) Amplitude and (b) intensity distributions in single-slit diffraction pattern.

$$\alpha = \frac{\pi a \theta}{\lambda}$$

by definition, then for the *n*th maximum, the value of $\pi a \theta_n / \lambda$ is constant. Thus $\theta_n \propto 1/a$, so that if we increase a, θ_n decreases, which means that the whole pattern closes in towards the centre. Similarly, if we decrease a, θ_n increases and the whole pattern expands outwards. This reciprocal relationship between size of slit and size of pattern is of considerable importance and interest in diffraction theory.

EXAMPLE 10.1 Diffraction of sound by an open window

Consider the following question: light and sound are both waves so why is it that light always travels in straight lines, but sound can go round a corner?

An appreciation of diffraction enables us to answer this. We have seen that it *is* possible for a ray of light to deviate from a straight line if it passes through an aperture, but this effect will not be noticeable if the aperture is large compared with the wavelength of light ($\sim 5 \times 10^{-7}$ m). If for instance a beam of light from a torch passes through a hole of radius 1 cm in a black card, it travels in a straight line and diffraction effects are negligible. But if for instance a sodium street lamp is viewed through an umbrella, an unexpected pattern is observed. This occurs because the umbrella material contains very many tiny holes of dimensions which are not large compared with the wavelength of light.

In the case of sound, the wavelengths are much greater (problem 4 in Chapter 4), and sounds, especially of low frequency (long wavelength), will readily travel round corners.

Consider a room with an open window of dimensions 1×1 m. Suppose that inside the room, well away from the window, is a sound source. We have then an acoustic analogue of the single slit already analysed. There will be significant diffraction of the sound so long as $(a \sin \theta)/\lambda \lesssim 1$ (equation (10.24)), whence $\lambda > 0.5$ m for an angle of $30°$. Since $c = 340 \text{ m s}^{-1}$ for sound, the corresponding frequency for significant diffraction of sound is $\lesssim 680$ Hz. A householder wishing to summon his dog from the garden should therefore shout towards the window – but would be ill adivsed to blow an ultrasonic dog whistle (unless there is an uninterrupted straight line between man and dog). ■

10.3.5 Fraunhofer diffraction and Fourier transforms

In the analysis in section 10.3.4, the diffraction pattern was viewed at the back focal plane of a converging lens. Such a pattern is known as a *Fraunhofer diffraction pattern* (FDP). In this case, the effect of the lens L_3 in Fig. 10.13 is to focus all parallel beams into fine lines; thus is recorded, at a finite distance, the diffraction pattern that would appear on an infinitely distant screen if the lens L_3 were absent. FDPs thus constitute a special case of diffraction which turns out to be much easier to analyse than the more general case, namely *Fresnel diffraction*, where the observation screen S_3 is in an arbitrary position. We shall not consider Fresnel diffraction further.

To illustrate an effectively one-dimensional case of Fraunhofer diffraction, we proceed as follows. Imagine rectangular coordinate axes drawn in the plane of the diffracting screen S_2 such that the origin is in the centre of the slit, the x-axis perpendicular to the slit and the y-axis parallel to the slit. Now, with these coordinate axes in place, imagine S_2 to be replaced by a screen whose transparency varies in the x-direction but does not vary in the y-direction. The transparency is a function, therefore, only of x; let us call it $g(x)$. The function $g(x)$ is such that those parts of the screen that are totally transparent have $g(x) = 1$ and those that are completely opaque have $g(x) = 0$. It is possible to imagine the screen being a film, parts of which could have transparencies between zero and unity. In such a case, the film would have to be of very high quality and, in particular, of very uniform thickness, so that no part of the film introduced a phase change to the incident radiation different from that suffered by any other part.

For diffraction at zero angle, the diffracted amplitude at the centre of the screen S_3 is $\int_{-\infty}^{\infty} g(x) \, dx$ (for light of unit incident amplitude) since each elementary part of the transparency (of width dx) is being added in phase to every other part. (We may put the integral limits formally at $-\infty$ and $+\infty$ on the understanding that $g(x)$ is zero for values of x outside the screen S_2.) But for diffraction at angle θ, we must add each elementary contribution with regard to its phase. If the phase difference between the contribution at $x = 0$ and that at an arbitrary value of x is ϕ_x, then $g(x)$ in the above integral must be replaced (in the complex exponential notation) by $g(x) \exp(-i\phi_x)$. By an obvious extension of the argument which led to (10.23), we have

$$\phi_x = \frac{2\pi}{\lambda} x \sin \theta.$$

Thus the total diffracted amplitude at angle θ is

$$\int_{-\infty}^{\infty} g(x) \exp(-2\pi i x \sin \theta/\lambda) \, dx \tag{10.27}$$

instead of the simpler $\int_{-\infty}^{\infty} g(x) \, dx$ to which (10.27) reduces for $\theta = 0$. We now put

$$s = \sin \theta/\lambda \tag{10.28}$$

so that the total diffracted amplitude at angle θ (corresponding to a given value of s), which we will call $G(s)$, is, from (10.27),

$$G(s) = \int_{-\infty}^{\infty} g(x) \exp(-2\pi i x s) \, dx. \tag{10.29}$$

Equation (10.29) is a result fundamental to Fraunhofer diffraction. We see, by comparison of (10.29) and (7.15), namely

$$G(f) = \int_{-\infty}^{\infty} g(t) \exp(-2\pi i f t) \, dt, \tag{7.15}$$

that $g(x)$ and $G(s)$ are Fourier-transform pairs, with x substituted for t and s for f. Here, therefore, we have a *spatial* rather than a temporal example of the application of the Fourier transform. In diffraction jargon, the space x is known as *real space* and the space s as *reciprocal space*. In our example, real space is along the x-axis on the screen S_2. Reciprocal space, on the other hand, is as follows. For small θ, for which $\sin \theta \cong \theta$, we have $s = \theta/\lambda$ from (10.28) and so, for this range of θ, reciprocal space is effectively along a line on S_3 parallel to the x-axis. For larger θ, however, we see a distorted picture in reciprocal space because θ and $\sin \theta$ become progressively more unequal. Nevertheless, for $0 \leqslant \theta \leqslant \pi/2$, there is a one-to-one relationship between a point on S_3 and a reciprocal–space point so that information in this space can still be retrieved.

The example of diffraction by a slit can easily be analysed as a special case. In x-space, the slit has a transparency of unity for $-a/2 \leqslant x \leqslant a/2$ and one of zero outside these limits. The transparency $g(x)$ is thus $\mathrm{rect}(x/a)$ as reference to problem 3 in Chapter 8 shows. The same problem also reveals that the FT of $\mathrm{rect}(x)$ is $\mathrm{sinc}(s)$ and the similarity theorem (8.4) immediately shows that the diffracted amplitude is

$$G(s) = a \, \mathrm{sinc}(as) \tag{10.30}$$

(the modulus sign of the similarity theorem not being needed because a is here positive). We see that we have recovered the result (10.26) for the amplitude of diffraction from the single slit.

A fundamental property of FDPs may be determined directly from the shift theorem. If the diffracting screen S_2 is moved a distance x_0 in the x-direction, the transparency is now

$$h(x) = g(x - x_0)$$

rather than the original $g(x)$. Its FT, by the shift theorem (8.2), is now

$$H(s) = \exp(-2\pi i x_0 s)\ G(s)$$

instead of the original $G(s)$. But we perceive not the amplitude $H(s)$ but only the intensity which is proportional to $|H(s)|^2$. But $|H(s)|^2 = |G(s)|^2$, since the factor $\exp(-2\pi i x_0 s)$ has a modulus of unity for all s. So, as the diffracting screen is moved along the x-direction, there is no perceivable change whatsoever occurring to its FDP. In other words, the FDP is completely insensitive to the absolute position of the diffracting screen.

We mention, as an aside, that S_2 may consist of two (or more) diffracting screens in contact with each other. If the transparencies of the screens are respectively $g_1(x)$ and $g_2(x)$, then a little thought will reveal that the overall transparency $g(x)$ is $g(x) = g_1(x)g_2(x)$. The diffracted amplitude $G(s)$ from such a composite screen is immediately given by the convolution theorem as $G(s) = G_1(s)*G_2(s)$, where G_1 and G_2 are the respective FTs of g_1 and g_2.

So far, we have considered only cases which are effectively one-dimensional. To obtain two-dimensional FDPs, the restricting slit S_1 in Fig. 10.13 must be replaced by a pinhole and the diffracting screen S_2 may now have a transparency which is a function of both x and y. Indeed, if $\mathbf{r}$ is the position vector on the screen S_2, we may think of the transparency as $g(\mathbf{r})$. The diffracted amplitude now also varies in two dimensions on the plane of S_3 and we may define a two-dimensional reciprocal space $\mathbf{s}$ with a diffracted amplitude $G(\mathbf{s})$. Although we shall not do the analysis, it turns out that

$$G(\mathbf{s}) = \int_{-\infty}^{\infty} g(\mathbf{r})\ \exp(-2\pi i \mathbf{r} \cdot \mathbf{s})\ d\mathbf{r},$$

where $d\mathbf{r}$ is an elementary area in $\mathbf{r}$ space. $G(\mathbf{s})$ is known as the *two-dimensional Fourier transform* of $g(\mathbf{r})$ and Fourier's theorem takes the form

$$g(\mathbf{r}) = \int_{-\infty}^{\infty} G(\mathbf{s})\ \exp(+2\pi i \mathbf{r} \cdot \mathbf{s})\ d\mathbf{s}$$

by analogy with (7.14) and (7.15). The shift theorem may readily be generalised so that the FT of $g(\mathbf{r} - \mathbf{r}_0)$ ($\mathbf{r}_0$ is a constant vector shift) is $\exp(-2\pi i \mathbf{r}_0 \cdot \mathbf{s})\ G(\mathbf{s})$ analogous to (8.2). The similarity theorem (8.4) may also be generalised so that the FT of $g(a\mathbf{r})$ (a being a constant non-zero scalar) is $(1/a^2)\ G(\mathbf{s}/a)$.

Fraunhofer diffraction has applications far wider than the two-dimensional FDP briefly treated above. If a solid is irradiated with X-rays, it diffracts them and the diffraction pattern is related to the three-dimensional Fourier transform $G(\mathbf{s})$ of the three-dimensional electric charge density $g(\mathbf{r})$ in the solid. If $g(\mathbf{r})$ can be determined, then one obtains knowledge of how the atoms are arranged in the solid; this is the basic aim of the crystallographer. The process is not straightforward because only the intensity pattern $|G(\mathbf{s})|^2$ may be observed, the phase of the complex amplitude $G(\mathbf{s})$ therefore not being determinable. Crystallographers have, over the years, applied much ingenuity to overcoming the phase problem and much success in the solution of evermore complex structures has been achieved with the application of Fourier-transform techniques to solids and their diffraction patterns.

10.4 The wave–particle duality

As a final section to this book, we present, very briefly, an introduction to the wave representation of atomic and sub-atomic particles. Throughout the history of physics until the end of the nineteenth century, waves and particles were considered as quite separate entities. It is, however, a remarkable fact of twentieth-century physics that these particles may be considered as waves, and indeed, in certain contexts, *must* be so considered. It is now known that material particles possess both particle-like and wave-like properties, one or the other of these being the more prominent depending on the experiment. The subject of this duality (quantum mechanics) erupted in the 1920s, notably at the hands of Louis de Broglie, Erwin Schrödinger, Paul Dirac and Werner Heisenberg, all of whom won physics Nobel Prizes.

A study of quantum mechanics is quite outside the scope of this book but a few comments on the implications of the duality will be made. In quantum mechanics, a particle is represented by a wave, the spatial component $\psi(x)$ of which is such that the probability of finding the particle between x and $x + dx$ is equal to $|\psi(x)|^2\, dx$. (The quantity $\psi(x)$ is usually complex and we are considering here only one spatial dimension.) If $\psi(x)$ is of the form of a wave group (see section 6.10), then the implication is that the particle is somewhere within the group with the probabilities given above. The wave describes our essential ignorance of where (even in principle) the particle is. The width, Δx, of the group describes the limitation of our knowledge of the position of the particle.

There are implications of the finite Δx consequent on the properties of Fourier transforms discussed in section 7.4 as we shall now see. In that section, we discussed the inequality

$$\Delta t\, \Delta f \geqslant 1/4\pi \qquad\qquad [7.17]$$

which says that the product of the width Δt of a signal and the width Δf of the FT of the same signal must exceed $1/4\pi$ (equality occurring if and only if the signal is Gaussian). For the spatial (rather than the temporal) version of this, we replace t by x and f by the spatial frequency σ (the *wavenumber*) which is the reciprocal of λ. We obtain

$$\Delta x\, \Delta\sigma \geqslant 1/4\pi. \qquad\qquad (10.31)$$

Inequality (10.31) says that if we have a wave function $\psi(x)$ of width Δx, then its spectrum has a width $\Delta\sigma$ such that the inequality is obeyed.

Now, it is the proposition of de Broglie (1924) that the momentum p of a particle is proportional to σ, the constant of proportionality being Planck's constant, h. That is to say,

$$p = h\sigma. \tag{10.32}$$

Differentiating (10.32), we obtain $\Delta p = h \, \Delta\sigma$, which, when put into equation (10.31), yields

$$\Delta x \, \Delta p \geqslant h/4\pi. \tag{10.33}$$

Inequality (10.33) is a statement of the *Heisenberg uncertainty principle* (1927). The inequality reveals fundamental limitations to the knowledge that we may acquire about a particle. Suppose that we know (at a given time) the value of Δx, perhaps because we have confined the particle to a box of this width. Then there is an essential uncertainty Δp in its momentum of at least $h/(4\pi \, \Delta x)$ by (10.33). It is not a question of experimental sophistication. The inequality states that the product of the two widths must be at least $h/4\pi$; no experimental technique can, even in principle, be devised that will lessen this product. This uncertainty, then, is a direct consequence of the nature of the relationship between x space and its reciprocal σ space. In other words, as soon as one represents a particle by a wave group $\psi(x)$ with a finite Δx, then a finite $\Delta\sigma$ (and hence Δp) obeying inequality (10.31) is inevitable by the properties connecting a signal with its Fourier transform. The more precisely we know the particle's position (i.e. the smaller Δx is), the greater is our ignorance of the momentum p because Δp must be large by (10.33).

Electrons were at first firmly considered as particles. The ratio e/m (charge to mass) was first determined by J.J. Thomson in 1897; e itself was first determined by Townsend and Thomson in about 1900 and in a famous experiment by R.A. Millikan in 1909 for which he received the 1923 physics Nobel Prize. So, evidently, the electron was a particle with known charge and mass. However, de Broglie's proposition $p = h\sigma$ directly linked a particle property (the momentum p) with a wave property (the wavenumber σ). In 1927, Davisson and Germer, in a search for wave-like properties of the electron, performed the celebrated experiment of directing a beam of electrons onto a nickel crystal; diffraction effects analogous to those described in the last section for X-ray irradiation of a crystal were observed. Thus the wave–particle duality was established; entities previously considered as particles were shown to have wave properties.

But could the opposite be true? Electromagnetic waves were at first thought of as waves. However, Planck's (1901) energy-quantisation postulate and Einstein's quantum theory of the photoelectric effect (1905) produced the proposition that electromagnetic radiation was propagated in quanta of energy, the energy E of each quantum being the product of Planck's constant and the frequency, i.e.

$$E = hf; \tag{10.34}$$

it was thus realised that these waves had particle properties.

Equation (10.34) is a direct relationship between a mechanical property (energy E) and a wave property (frequency f) in a way totally analogous to that proposed by de Broglie connecting momentum with wavenumber. The electromagnetic quantum is known as the photon. Thus, the modern view is that waves and particles are but different aspects of the same thing.

Problems

1. In the SI system of units, μ_0 is *defined* as exactly $4\pi \times 10^{-7}$ H m^{-1}, and ε_0 is a measured quantity, which has a value 8.85×10^{-12} F m^{-1}. Show that the speed of electromagnetic waves in free space is 3.00×10^8 m s^{-1}. The reader familiar with electromagnetism is challenged to show that the dimensions of (H m^{-1} F m^{-1})$^{-1/2}$ are indeed [m s^{-1}].

2. The refractive index of water, measured by various optical methods, is approximately 1.3. The relative permittivity, measured by electrical methods, is approximately 80. Explain why there is an apparent violation of equation (10.13).

3. This problem is recommended to a reader wishing to obtain further insight into 'circular polarisation' which is important in optics.

A string is stretched along the z-axis of coordinates, and a sinusoidal transverse wave $x = a \sin(\omega t - kz)$ is propagated along it. A second sinusoidal wave $y = a \sin(\omega t - kz)$ is now superimposed, its plane being perpendicular to that of the first. By making sketches of the disturbance in the x–y plane at $z = 0$, show that the resulting wave is of amplitude $\sqrt{2}\, a$, polarised in a plane at $45°$ to the x-axis.

Suppose now that the second wave is instead $a \cos(\omega t - kz)$. Show that an instantaneous photograph of the string would reveal it to have the shape of a helix. Would it be a right-handed or a left-handed helix? (Compare it with a conventional screw thread.) Consider an observer at a fixed position z_0. Show that in this z_0 plane, he would observe a disturbance rotating in a circle. If he is *facing* the incoming waves, would the disturbance rotate clockwise or anticlockwise?

4. A sound source at ground level is detected by a distant observer also at ground level. There is a wind in the direction from the source to the observer. Remembering that the wind speed will increase with height (why?), construct Huygens wave fronts from the source. Hence explain why 'sound travels better with the wind'.

5. Prove the result stated in section 10.3.3 for Young's slits, namely that the bright fringes produced on a screen are $\lambda D/a$ apart. If a thin piece of glass were placed over one of the slits on the side nearer the screen, what would be the effect on the fringe pattern?

6. Use a calculator to plot the functions $(\sin \alpha)/\alpha$, and $(\sin^2 \alpha)/\alpha^2$. Hence find the first value of α (other than zero) for which $(\sin^2 \alpha)/\alpha^2$ is a maximum, and the value of that maximum.

7. A diffraction grating is made in such a way that its transparency $g(x)$ varies sinusoidally along its length. What would be the appearance of the Fraunhofer diffraction pattern? If $g(x)$ were instead a Gaussian function $g_0 \exp(-x^2/a^2)$, what would now be the appearance of the pattern?

Appendix
Solution of the differential equations for free, damped and forced vibrations

A.1 Undamped vibrations

The differential equation of motion of a particle moving in one dimension under the influence of a restoring force proportional to the displacement was introduced in Chapter 2 (2.1), namely

$$\frac{d^2x}{dt^2} + \omega^2 x = 0. \tag{A.1}$$

Here ω is a real and positive constant, and $\omega^2 x$ is the acceleration towards the origin.

To solve A.1, we first multiply throughout by $2\,(dx/dt)$,

$$2\frac{dx}{dt}\frac{d^2x}{dt^2} + 2\omega^2 x \frac{dx}{dt} = 0.$$

Integration with respect to t gives

$$\left(\frac{dx}{dt}\right)^2 + \omega^2 x^2 = \text{constant}.$$

As will become evident, the most convenient way of expressing this constant is $\omega^2 a^2$, where a is another constant. We are anticipating the result that a will be shown to be the amplitude of the motion. Hence,

$$\frac{dx}{dt} = \omega(a^2 - x^2)^{1/2}. \tag{A.2}$$

This expression gives the velocity of the particle as a function of x.

To obtain x as a function of t, we rearrange (A.2) and integrate again,

$$\int \frac{dx}{(a^2 - x^2)^{1/2}} = \omega \int dt$$

whence $\sin^{-1}(x/a) = \omega t + \varepsilon$, where ε is another constant of integration. Thus,

$$x = a \sin(\omega t + \varepsilon).$$ (A.3)

A.2 Damped vibrations

We now solve (3.8):

$$\frac{d^2 x}{dt^2} + 2b \frac{dx}{dt} + \omega_0^2 x = 0.$$ (A.4)

This equation is appropriate, for instance, to a single particle of mass m subject to a damping force $m(2b\, dx/dt)$ and a restoring force $m\omega_0^2 x$. Here ω_0 must not be interpreted as the circular frequency of any ensuing damped oscillations, but rather as the circular frequency of the oscillations which would have occurred had there been no damping.

(A.4) is an example of a *second-order* differential equation, which means that the highest derivative involved in the equation is the second, $d^2 x/dt^2$. Now it can be shown that the general solution of such an equation will have *two* arbitrary constants. For instance, the solution of (A.1) has been shown to be (A.3), where a and ε are the two arbitrary constants. To take another simple example, the solution of the equation $d^2 y/dx^2 = 1$ is readily shown to be $y = \frac{1}{2}x^2 + Ax + B$, where A and B are the two arbitrary constants.

The converse is also true. Any solution of a second-order differential equation with two arbitrary constants is a general solution. Therefore, to solve (A.4) it is merely necessary to find any solution involving two arbitrary constants. It would therefore have been quite in order to have written down the solution (A.11) below, and verified that it satisfies (A.4). However, for the sake of elegance, we proceed forwards from (A.4) to the solution.

(A.4) can be written

$$\left(\frac{d}{dt} + \alpha_1 \right)\left(\frac{dx}{dt} + \alpha_2 x \right) = 0$$ (A.5)

where $\alpha_1 + \alpha_2 = 2b$, and $\alpha_1 \alpha_2 = \omega_0^2$. Elimination of α_2 from these two latter equations gives

$$\alpha_1^2 - 2b\alpha_1 + \omega_0^2 = 0.$$

Hence

$$\alpha_1 = b \pm \sqrt{(b^2 - \omega_0^2)}$$

and

$$\alpha_2 = b \mp \sqrt{(b^2 - \omega_0^2)}.$$ (A.6)

With the substitution

$$u = \frac{dx}{dt} + \alpha_2 x \tag{A.7}$$

(A.5) becomes $du/dt + \alpha_1 u = 0$. Rearranging, $du/u = -\alpha_1 dt$, whence on integration, $\log_e u = -\alpha_1 t + \text{constant}$. Therefore, $u = K \exp(-\alpha_1 t)$, where K is an arbitrary constant. From (A.7),

$$\frac{dx}{dt} + \alpha_2 x = K \exp(-\alpha_1 t). \tag{A.8}$$

We have reduced the problem to that of solving a first-order differential equation. The reader familiar with the solution of such equations will recognise that the procedure is to multiply throughout by the *integrating factor*, which in this case is $\exp(\alpha_2 t)$, giving

$$\frac{dx}{dt} \exp(\alpha_2 t) + \alpha_2 \exp(\alpha_2 t)x = K \exp(\alpha_2 - \alpha_1)t.$$

The left-hand side is the derivative of $x \exp(\alpha_2 t)$, and so

$$x \exp(\alpha_2 t) = K \int \exp(\alpha_2 - \alpha_1)t \; dt$$

$$= \frac{K}{\alpha_2 - \alpha_1} \exp(\alpha_2 - \alpha_1)t + B, \tag{A.9}$$

where B is a constant. We have assumed that $\alpha_1 \neq \alpha_2$ here. The case where $\alpha_1 = \alpha_2$ will be considered in the next section. Writing A for $K/(\alpha_2 - \alpha_1)$, we have

$$x = A \exp(-\alpha_1 t) + B \exp(-\alpha_2 t). \tag{A.10}$$

Finally, substitution from (A.6) gives the final solution

$$x = A \exp[-b + \sqrt{(b^2 - \omega_0^2)}]t + B \exp[-b - \sqrt{(b^2 - \omega_0^2)}]t. \tag{A.11}$$

(The lower signs of $\pm$ and $\mp$ in (A.6) have been used. Clearly it makes no difference which choice is adopted.)

A.3 Examination of the solution for damped vibrations

The behaviour of x as a function of t depends on the size of the damping constant b, and it is necessary to examine separately three different cases.

1. *Case 1, $b > \omega_0$ (heavy damping)*. In this case, α_1 and α_2 are real quantities, and x is interpreted from (A.10) and (A.11) in a straightforward manner. It is simply the sum of two functions, exponentially decreasing at different rates.

2. *Case 2, $b = \omega_0$ (critical damping)*. Here $\alpha_1 = \alpha_2 = b$. It will be recalled, however, that in the derivation of (A.11), it was assumed that α_1 and α_2 were different. We

therefore return to (A.9), which becomes on substituting b for α_2, and writing A instead of K,

$$x \, e^{bt} = A \int dt = At + B$$

where B is a constant of integration. The solution is therefore

$$x = (At + B) \, e^{-bt}. \tag{A.12}$$

3. *Case 3, $b < \omega_0$ (light damping).* The solution (A.11) is valid in this case, but the exponents are complex quantities. We therefore rewrite (A.11) in the form

$$x = e^{-bt}[A \exp i\sqrt{(\omega_0^2 - b^2)}t + B \exp - i\sqrt{(\omega_0^2 - b^2)}t]$$
$$= e^{-bt}[(A + B) \cos \sqrt{(\omega_0^2 - b^2)}t + i(A - B) \sin \sqrt{(\omega_0^2 - b^2)}].$$

For generality, A and B must be complex quantities, but clearly $(A + B)$ and $i(A - B)$ must both be real. Putting these respectively as D and C, we have

$$x = e^{-bt}[C \sin \sqrt{(\omega_0^2 - b^2)}t + D \cos \sqrt{(\omega_0^2 - b^2)}t], \tag{A.13}$$

which is one way of expressing the general solution.

An alternative form of the solution is obtained by introducing the phase constant ε,

$$x = a \, e^{-bt} \sin[\sqrt{(\omega_0^2 - b^2)}t + \varepsilon] \tag{A.14}$$

where a is interpreted as the amplitude of the motion at time $t = 0$. This solution is conveniently expressed as

$$x = a \, e^{-bt} \sin(\omega t + \varepsilon) \tag{A.15}$$

where

$$\omega = \sqrt{(\omega_0^2 - b^2)}. \tag{A.16}$$

A.4 Forced vibrations

This section is concerned with the problem of a damped oscillator subjected to a sinusoidally varying force. The equation of motion is (compare with (A.4))

$$\frac{d^2x}{dt^2} + 2b\frac{dx}{dt} + \omega_0^2 x = P \sin pt. \tag{A.17}$$

This equation would appear to be rather formidable, but there is a simple way of proceeding. Suppose we were in our investigations to come across *some particular solution* x_P (i.e. one in which there were no arbitrary constants). Then

$$\frac{d^2x_P}{dt^2} + 2b\frac{dx_P}{dt} + \omega_0^2 x_P = P \sin pt.$$

Subtracting this from (A.17), and writing x_C for $x - x_P$,

$$\frac{d^2 x_C}{dt^2} + 2b \frac{dx_C}{dt} + \omega_0^2 x_C = 0.$$

This will be recognised as an equation which has been solved in its generality in section A.2.

The solution x of (A.17) is therefore the sum of two parts (a) *any* particular solution (or particular integral) x_P of this equation and (b) the solution (or complementary function) x_C of the equation obtained by replacing the right-hand side of (A.17) by zero. This complete solution is a general solution, because the complementary function x_C contains two arbitrary constants, and (A.17) is a second-order equation.

The complementary function is readily found by the analysis given in section A.2, and it merely remains to seek a particular solution to (A.17).

Now it will be recalled that the complementary function is one which dies away as time progresses, therefore the steady-state oscillations will be described in the particular integral. These will have the same frequency as the driving force. It is therefore clear that we should try a solution of the form

$$x = F \sin(pt - \delta) \tag{A.18}$$

where F is the steady-state amplitude, and δ is the phase lag of the oscillator behind the driving force. (The use of a negative sign in (A.18) will ensure that δ is always a quantity lying between 0 and π.)

Substitution of (A.18) in (A.17) gives

$$(\omega_0^2 - p^2)F \sin(pt - \delta) + 2bpF \cos(pt - \delta) = P \sin pt.$$

The only way in which this equation can be satisfied at *all* times t is for the coefficients of sin pt *and* cos pt to be equal on both sides of the equation, whence

$$(\omega_0^2 - p^2)F \cos \delta + 2bpF \sin \delta = P \tag{A.19a}$$

and

$$-(\omega_0^2 - p^2)F \sin \delta + 2bpF \cos \delta = 0. \tag{A.19b}$$

The second of these equations gives the phase,

$$\tan \delta = \frac{2bp}{\omega_0^2 - p^2}. \tag{A.20}$$

F is now obtained from (A.19a) as follows

$$(\omega_0^2 - p^2)F + 2bpF \tan \delta = P \sec \delta = P(1 + \tan^2 \delta)^{1/2}.$$

Hence, from (A.20),

$$\left[(\omega_0^2 - p^2) + \frac{4b^2 p^2}{\omega_0^2 - p^2} \right] F = P \left[\frac{(\omega_0^2 - p^2)^2 + 4b^2 p^2}{(\omega_0^2 - p^2)^2} \right]^{1/2}$$

Rearrangement gives finally

$$F = \frac{P}{[(\omega_0^2 - p^2)^2 + 4b^2 p^2]^{1/2}}.$$

The particular integral is therefore given by

$$x = \frac{P \, \sin(pt - \delta)}{[(\omega_0^2 - p^2)^2 + 4b^2 p^2]^{1/2}} \tag{A.21}$$

where the phase constant δ is given by (A.20).

The complete solution of the original equation (A.17) is therefore the sum of the right-hand side of (A.21) and that of (A.11), (A.12) or (A.14) as appropriate.

Answers to problems

Chapter 2

(2.1) $\alpha = a \cos \varepsilon$, $\beta = a \sin \varepsilon$; (ii) $a = (\alpha^2 + \beta^2)^{1/2}$, $\varepsilon = \tan^{-1} \beta/\alpha$. 5; a is the hypotenuse of a right-angled triangle where the other sides are 3 and 4.

(2.2) (i) 10^{-4} J; (ii) 1.0001 N.

(2.3) $2\pi(l/2g)^{1/2}$.

(2.5) (i) 2; (ii) 3. 25 and 26 cycles of x and y respectively.

(2.7) $2\pi[Mm/k(M + m)]^{1/2}$.

(2.8) $(1/2\pi)\,(k/2m)^{1/2}$, in phase, the lower having double the amplitude of the upper; $(1/2\pi)\,(2k/m)^{1/2}$, in antiphase, with the same amplitudes.

$$\frac{u}{3}\sqrt{\left(\frac{2m}{k}\right)}\left[\sin\sqrt{\left(\frac{k}{2m}\right)}t + \sin\sqrt{\left(\frac{2k}{m}\right)}t\right]$$

(2.9) $(3/2\pi)\sqrt{(T/ml)}$, the particles all in phase, the outer ones having the same amplitude, and the middle one double this amplitude; $(3/2\pi)\sqrt{(3T/ml)}$, the middle particle stationary, the outer two being in antiphase and having the same amplitude; $(3/\pi)\sqrt{(T/ml)}$, the amplitudes all being the same, with the two outer ones in phase, and the middle one in antiphase with them.

Chapter 3

(3.1) Less; 0.34 mm.

(3.5) (i) $(\log_e 2)/300$ s^{-1}; (ii) $300/\log_e 2$ s, (iii) 4.6×10^{-3}; (iv) 684; (v) 1.5×10^{-5} J. Of the order of 10^{-3} Hz.

(3.8) $R \geqslant 2000 \ \Omega$. (i) 1.59 kHz, (ii) 2 ms, (iii) 10, (iv) $10^{-6} \pi$ J.

(3.9) (i) $\sqrt{99}/2\pi$ MHz; (ii) $\sqrt{98}/2\pi$ MHz, $5/\pi$ MHz; (iii) 5.

Chapter 4

(4.1) (i) 4/3 in the positive direction of x; (iv) ω/k in the negative direction of x; (v) 1/2 in the positive direction of x; (viii) 2 in the negative direction of x.

(4.2) 25, 5, 5, $2\pi/5$.

(4.3) $2\pi f/c$ dx. Q lags behind P.
(4.4) 11.3 m, 22.7 mm.
(4.5) (i) 3 m; (ii) 6×10^6 m; (iii) 6×10^{14} Hz; (iv) 3×10^{18} Hz.
(4.8) k_1 and k_2 are the components of **k** considered as a vector; $\omega(k_1^2 + k_2^2)^{-1/2}$.

Chapter 5

(5.2) (i) $2A/3$; (ii) $-A/3$. There is a phase change.
(5.3) Acoustic impedance matching.
(5.4) $0.533\ \mu\text{H m}^{-1}$, $83.3\ \text{pF m}^{-1}$.
(5.5) 5/9 absorbed, 4/9 reflected; 5.

Chapter 6

(6.1) $34.3\ \text{m s}^{-1}$, 81.
(6.2) $2\sqrt{(l/g)}$.
(6.4) $774\ \text{m s}^{-1}$.
(6.5) $2.19\ \text{N m}^{-2}$.
(6.6) (i) $2.40 \times 10^9\ \text{N m}^{-2}$; (ii) about 3% greater than its surface value.
(6.7) $5.06\ \text{km s}^{-1}$.
(6.8) $Kf/(c^2\rho C_V) = 2.3 \times 10^{-7}$.
(6.9) (i) 0.5 m; (ii) 0.264 m.
(6.10) $108\ \Omega$.
(6.11) The car is moving at $37.8\ \text{m s}^{-1}$; therefore, the driver is exceeding the speed limit.
(6.13) $f_{O+S} = f(c + v_W - v_O)/(c + v_W - v_S)$.
(6.15) Of the order of 6×10^{-12} m.
(6.17) $3.95\ \text{m s}^{-1}$.
(6.18) No. Same as the wave velocity.
(6.19) $\lambda \ll 1.7$ cm, $0.68\ \text{m s}^{-1}$.
(6.20) $c = V + K\lambda$, where K is a constant. $K = 0$ for a physically acceptable solution.

Chapter 7

(7.2) $A_0 = h/\pi$. For $n \neq 0$, $A_n = \{2h/[\pi(1 - n^2)]\}\cos(\pi n/2)$.
(7.4) If the function has been shifted by an integral number of periods, it is then indistinguishable from the unshifted function and so, in this case, $H_n = G_n$.
(7.5) The phase of the transform at frequency $-f$ is equal and opposite to that for $+f$.
(7.6) Required FT is $\pi\tau\exp(-2\pi\tau\,|f|)$.
(7.7) The FT of the signal is a sinc function centred on 5000 Hz. With the fourth zero above this frequency of the function chosen as the bandwidth boundary, the required bandwidth is from 4000 to 6000 Hz.

Chapter 8

(8.3) There is zero amplitude at frequencies 1, 2, 3, …

(8.6) $10f_1$ to $10f_2$.

(8.7) The required FTs are respectively $(\pi/\alpha)^{1/2} \exp(-\pi^2 f^2/\alpha)$, $-if (\pi/\alpha)^{3/2} \exp(-\pi^2 f^2/\alpha)$ and $(\pi/4\alpha^3)[1 - (2\pi^2 f^2/\alpha)] \exp(-\pi^2 f^2/\alpha)$.

(8.8) $i(t) = [v_0/(R^2 + X^2)^{1/2}] \cos(2\pi f_0 t - \phi)$, where $\phi = \tan^{-1}(X/R)$.

(8.10) The frequency components which will not reach the loudspeaker are at frequencies n/T.
       Maximum width of slit is 0.027 mm.

(8.11) (i) Zero and $2f_0$; (ii) for $f_2 > f_1$, frequencies are zero, $f_2 - f_1$, $2f_1$, $f_1 + f_2$, $2f_2$.

Chapter 9

(9.1) 17. 96.2 cents.

(9.2) $nc/2L$.

(9.3) 12. Twelve perfect fifths are very close to seven octaves. This is one reason for having 12 notes in the musical scale.
       Four perfect fifths (minus two octaves) are very close to the simple ratio of 5/4 (a major third). Thus, for example, C and E produce a consonant sound.

(9.4) Lower.

(9.5) (i) 100 Hz; (ii) 300 Hz.

(9.7) $y(t) = \sum_n A_n \sin(n\pi x/L) \cos(n\pi ct/L)$

       where $A_n = (16h/n^2\pi^2) [\sin(n\pi/4) - \sin(3n\pi/4)]$ and $c = \sqrt{(T/\mu)}$.

       The missing harmonics are 1, 3, 4, 5, 7, 8, 9, ….

(9.8) $(16h^2 T/\pi^2 n^2 L) \sin^2(\pi n/2)$.

(9.9) 10^4 km! No account is taken of absorption by the air.

(9.10) $p_1 \sin[2\pi \times 361 \exp(0.5545t)]$.

(9.11) 1.43×10^{-2} N m^{-2}; 1.40×10^{-2} N m^{-2}.

(9.12) 56.2 dB.

(9.13) 6 dB.

(9.14) If the control were 'linear', a small turn of the control would produce a sudden increase in loudness at the low-intensity range and hardly any effect at all at the high-intensity range.

Chapter 10

(10.2) The measurements are made at very different frequencies.

(10.3) Right-handed. Clockwise. (Remember that x, y, z must form a right-handed coordinate system.)

(10.5) The centre of the pattern moves.

(10.6) 4.49, 0.047.

(10.7) There are three lines, one being at $\theta = 0$, the others being equally spaced either side of this.
       The intensity of the pattern is
       $$I = I_0 \exp(-2\pi^2 a^2 \sin^2 \theta/\lambda^2)$$
       where I_0 is a constant.

Bibliography

General

G. R. Baldock and T. Bridgeman, *The Mathematical Theory of Wave Motion*. Ellis Horwood: Chichester 1981. Not for the beginner. More advanced standard than this book.

H. J. J. Braddick, *Vibrations, Waves and Diffraction*. McGraw – Hill: London 1965. General coverage.

W. C. Elmore and M. A. Heald, *The Physics of Waves*. McGraw-Hill: London 1969. A comprehensive, readable and informative book, of considerably more advanced standard than this one.

A. P. French, *Waves and Vibrations*. Van Nostrand Reinhold: Wokingham 1987.

I. G. Main, *Vibrations and Waves in Physics*. Cambridge University Press: Cambridge 1984. Contains chapter on water waves.

H. J. Pain, *The Physics of Vibrations and Waves*. John Wiley and Sons: Chichester 1976. Contains chapter on non-linear oscillations including shock waves.

A. B. Pippard, *The Physics of Vibrations*, two volumes. Cambridge University Press: Cambridge 1978. Excellent for the expert – new approaches to familiar topics – but not for the beginner.

Water waves

G. D. Crapper, *Introduction to Water Waves*. Ellis Horwood: Chichester 1984.

Solitons

P. J. Drazin and R. S. Johnson, *Solitons: An introduction*. Cambridge University Press: Cambridge 1989. Starts with water waves and becomes more general.

Chaotic vibrations

G. L. Baker and J. P. Gollub, *Chaotic Dynamics: An introduction*. Cambridge University Press: Cambridge 1990. Sets up the subject in a rigorous manner; mathematically demanding.

J. P. Crutchfield, J. Doyne Farmer, N. H. Packard and R. S. Shaw, 'Chaos'. *Scientific American*, December, 1986. Very readable review article.

J. Gleick, *Chaos: Making a new science*. Abacus: London 1993. Good general coverage, not heavily mathematical.

Piezoelectric effect

A. P. Crackell, *Ultrasonics*. Wykeham Publications: London 1980. Good general coverage.
J. R. Hook and H. E. Hall, *Solid State Physics*, 2nd edn. John Wiley and Sons: Chichester 1991. Student text with coverage of piezoelectricity.
R. Truell, C. Elbaum and B. B. Chick, *Ultrasonics Methods in Solid State Physics*. Academic Press: New York 1969. More a researcher's reference book – good on theory and experiment.

Fourier methods

R. Bracewell, *The Fourier Transform and its Applications*. McGraw-Hill: New York 1978. This is the classic book on the subject for the physicist and engineer. Mathematically sound, without being too 'purist'. A wide variety of applications is discussed, illustrated by a wealth of illuminating diagrams.
R. F. Hoskins, *Generalised Functions*. Ellis Horwood: Chichester 1979. Delta functions are examples of *generalised* functions, which are introduced in this book in a very good and fairly easy-to-read exposition.
E. G. Steward, *Fourier Optics: An introduction*. Ellis Horwood: Chichester 1983. Complements the chapters on Fourier methods in this book; includes chapters on optical imaging and interferometry.

Acoustics

W. E. Seto, *Theory and Problems of Acoustics*. Schaum's Outline Series, McGraw-Hill: New York 1971. A broad tour through the facts of acoustics in the unmistakable Schaum style. Very informative.
L. E. Kinsler and A. R. Frey, *Fundamentals of Acoustics*, 3rd edn. John Wiley and Sons: New York 1982. A classic book on the subject, with leanings towards applications such as architectural and underwater acoustics.

Optics

E. Hecht and A. Zajac, *Optics*. Addison-Wesley: London 1974. General coverage.

Wave mechanics

R. M. Eisberg and R. Resnick. *Quantum Physics of Atoms, Molecules, Solids, Nuclei and Particles*. John Wiley and Sons: New York 1974.

Index

Christopher Glennie
Prentice Hall Europe
Campus 400
Maylands Avenue
Hemel Hempstead
Herts., HP2 7EZ
UK

Please send me one copy of the *Waves Demonstration* disk free of charge.

Name _____

Address _____

_____

_____